ORGANIC AGRICULTURE IN THE U.S.

ORGANIC AGRICULTURE IN THE U.S.

ALISON J. WELLSON
EDITOR

Nova Science Publishers, Inc.
New York

For permission to use material from this book please contact us:
Telephone 631-231-7269; Fax 631-231-8175
Web Site: http://www.novapublishers.com

Library of Congress Cataloging-in-Publication Data

Organic agriculture in the U.S. / Alison J. Wellson (editor).
p. cm.
Includes index.
ISBN 13: 978-1-60021-305-2
ISBN 10: 1-60021-305-7
1. Organic farming--United States. 2. Ecosystem management--United States. 3. Organic farming--Economic aspects--United States. 4. Organic farming--Environmental aspects--United States. I. Wellson, Alison J.
S605.5.O725 2006
631.5'840973--dc22 2006019482

Published by Nova Science Publishers, Inc. ✦New York

CONTENTS

PREFACE

Organic farming is a form of agriculture that relies on ecosystem management and attempts to reduce or eliminate external agricultural inputs, especially synthetic ones. It is a holistic production management system that promotes and enhances agro-ecosystem health, including biodiversity, biological cycles, and soil biological activity.

In preference to the use of off-farm inputs, organic farming emphasizes management practices, taking into account that regional conditions require locally adapted systems. Utilizing both traditional and scientific knowledge, organic agricultural systems rely on agronomic, biological, and mechanical methods (these may require external inputs of nonrenewable resources, like tractor fuel), as opposed to using synthetic materials, to fulfill any specific function within the system. Organic farming is also associated with support for principles beyond cultural practices, such as fair trade and environmental stewardship. This book dives into cutting edge research within relation to this field. This includes frequently asked questions, price premiums, economic labeling, market expansions, Government-faciliated programs, growth patterns and publications from the United States Department of Agriculture.

In: Organic Agriculture in the U.S.
Editor: Alison J. Wellson, pp. 1-53
ISBN: 978-1-60021-305-2

Chapter 1

ORGANIC AGRICULTURE IN THE U.S.: PROGRAM AND POLICY ISSUES*

Jean M. Rawson

SUMMARY

Congress passed the Organic Foods Production Act (OFPA) in 1990 as part of a larger law governing U.S. Department of Agriculture (USDA) programs from 1990 through 1996 (P.L. 101-624, the Food, Agriculture, Conservation, and Trade Act of 1990). The act authorized the creation of a National Organic Program within USDA to establish standards for producers and processors of organic foods, and permit such operations to label their products with a "USDA Organic" seal after being officially certified by USDA-accredited agents. The purpose of the program, which was implemented in October 2002, is to give consumers confidence in the legitimacy of products sold as organic, permit legal action against those who use the term fraudulently, increase the supply and variety of available organic products, and facilitate international trade in organic products.

Due to the newness of the National Organic Program, it is difficult to gauge its overall impact on the organic industry. However, a recent USDA report states that the number of people seeking USDA accreditation to become certification agents increased by more than 130% between 2002 and 2005, suggesting significant growth in the number of farm and processing operations going into organic production. USDA estimates that sales of organic foods rose from $6 billion in 2000 to $10.4 billion in 2003. The annual rate of market growth since 1990 has remained steady at about 20%.

---

* Excerpted from CRS Report RL31595, Updated November 3, 2005

Policy issues affecting the National Organic Program since implementation largely reflect the differences in interpretation among stakeholders of the language and intent of OFPA and the actual operation of the program under the final rule. The most recent issues concern, first, USDA's efforts to write a new regulation governing livestock access to pasture, and second, a lawsuit brought by a plaintiff who argued that USDA's administration of the NOP has been too lenient concerning the use of certain synthetic ingredients in foods bearing the "USDA Organic" label, and too lenient concerning the process for converting a dairy herd from conventional to organic production. The final judgment in the case, which was issued in June 2005, requires the Department to develop new rules based on the original statutory language. The revised rules would become effective two years after the date of the judgement, up until which time products made under the old rules would be allowed into commerce.

Subsequently, however, conferees on the FY2006 USDA appropriations bill attached a provision that amends the OFPA in such a way as to largely maintain the NOP regulations as they were before the court decision. This action is likely to worsen the rift between those who welcomed the court ruling because they felt it would force USDA to write stricter regulations that would more closely reflect the statute, and those (a sizeable consortium of major food companies) who predicted that new rules written to satisfy the court decision might severely limit current and future lines of organic products and economically harm all sectors of the industry.

BACKGROUND

Organic farming, as defined in the final rule establishing the USDA National Organic Program (NOP), is "a production system that is managed in accordance with the [Organic Foods Production] Act and regulations ... to respond to site-specific conditions by integrating cultural, biological, and mechanical practices that foster cycling of resources, promote ecological balance, and conserve biodiversity."[1] This definition indicates that organic agriculture is both an approach to food production based on biological methods that avoid the use of synthetic crop or livestock production inputs (spelled out in detail in the December 2000 rule), and a broadly defined philosophical approach to farming that puts value on resource efficiency and ecological harmony.

Interest in organic farming migrated from Europe to the United States in the early 1900s. Beginning in the 1950s, as the U.S. public became more concerned about the potential adverse environmental and public health effects of agricultural chemicals and so-called "factory farming" methods, private research organizations began to conduct scientific investigations into non-chemical and non-intensive farming techniques, and a small but slowly increasing number of farmers began to adopt organic production practices. Except for a brief period from about 1978 to 1981, USDA did not conduct any activities in support of organic agriculture until OFPA required the Department to begin rulemaking to establish the National Organic Program in 1990.

ORGANIC SECTOR STATISTICS[2]

In 2001 (the most recent year for which data are available), there were 2.3 million acres of cropland and pasture under organic management in 48 states, with half of that amount having been added between 1997 and 2001. Nationwide there were about 7,000 certified organic operations in 2001, according to USDA. Fresh produce is the largest sector of the organic industry, with California, Washington, and Colorado having the greatest number of acres devoted to organic fruit and vegetable production. Colorado, Texas, and Montana have the most acreage of organic pasture for livestock. In the Northeast, Southeast, and Upper Midwest, small-scale growers of organic fruits, vegetables, herbs, and flowers are a significant component of individual states' agriculture industries.[3]

Since 1990, sales of organic products have been growing 20% per year on average. Estimated growth rates from 2005 through 2010 range from 9% to 16% annually. The estimated value of sales of organic foods was $6 billion in 2000 and $10.4 billion in 2003. In 2003, 47% of organic foods were sold through conventional retailers, 44% through natural food stores, and 9% through farmers' markets, restaurants, exports, and other marketing channels.[4] Despite sustained market growth, organic food sales constitute about 1.8% of the total U.S. retail sales of food. Various sources of export data estimated U.S. exports of organic foods at between $125 million and $300 million in the 2000-2002 period.[5] The biggest export market is Canada; other major markets are Japan, the European Union, and other countries in Asia.

The high growth rate in sales of organic products can be attributed in part to the higher prices that organic producers and processors receive for their products, according to USDA economists. They speculate that part of the price premium may be due to higher production costs and to higher demand relative to supply. For the 2000-2004 period, the annual average farmgate price premiums for fresh organic broccoli and carrots fluctuated from 75% to 133% above the prices of conventionally grown broccoli and carrots, according to ERS. Price premiums for the two vegetables at the wholesale level never went below 125% in the same period.[6] The ERS study goes on to say that:

> Laws of supply and demand, however, make it unlikely that price premiums contributing to higher profits and market growth can coexist over the long run: as long as higher profits exist, new suppliers will enter the market, and once market supply increases faster than demand, price premiums and the commensurate level of higher profits are likely to decline.... Many organic industry participants and observers believe that the price premiums ... need to decrease if organic foods are to penetrate much beyond the 2- to 3-percent level into the mainstream.[7]

The consumer studies that ERS reviewed did not permit any clear estimate of future demand for organic products. The studies showed that price, size, packaging, appearance, and concerns about health and nutrition, taste, food safety, and the environment all play varying roles in consumer decisions to buy organic food. Surveys on race, ethnicity, and income levels showed significant diversity.[8]

THE ORGANIC FOODS PRODUCTION ACT OF 1990

Congress passed the Organic Foods Production Act (OFPA) of 1990 (Title 21 of P.L. 101-624, the Food, Agriculture, Conservation, and Trade Act of 1990; the 1990 farm bill) with widespread support from organic industry groups, the National Association of State Departments of Agriculture, and other farm and consumer groups. The organic industry petitioned Congress to draft the act in the late 1980s, after it had been frustrated in its attempts to come to an internal consensus on production and certification standards. The industry maintained that federal standards would reduce consumer confusion over the many different state and private standards then in use, and would promote confidence in the integrity of organic products over the long term. Manufacturers of multi-ingredient organic food products stated that uniform standards would facilitate labeling. Others held that regulations would help the organic industry expand product lines and increase marketing opportunities. Industry analysts asserted that a consistent U.S. organic standard would facilitate access to a potentially lucrative international organic market.

The Organic Foods Production Act of 1990 authorized a National Organic Program to be administered by USDA's Agricultural Marketing Service (AMS). The act established a 15-member National Organic Standards Board (NOSB) to "assist in the development of standards for substances to be used in organic production" (referred to as the "National List") and to "provide recommendations to the Secretary regarding implementation."

Under the program, producers, processors and handlers who wish to market their products as organic are required to follow production practices as spelled out in detail in regulations (7 CFR 205). USDA accredits private and state certification agents, who visit producers, processors, and handlers to certify that their operations abide by the standards; they conduct annual reviews to verify continued compliance. It is illegal for anyone to use the word "organic" on a product if it does not meet the standards set in the law and regulations.[9] The presence of the "USDA Organic" seal on a product means it is 95% or more organic. Labels on products having 70% to 95% organic content can say "made with organic (specified ingredients or food groups)," but cannot carry the seal. Foreign organic producers and handlers wishing to export products to the United States may be certified by a USDA-accredited certification agent in their own country, if there is one; or USDA may accept certification by agents accredited by a foreign government; or, USDA may negotiate an equivalency agreement with another nation's organic program.[10]

The regulations under the OFPA are intended to set uniform *minimum* standards for organic production. States may adopt additional requirements after review and approval by USDA. Furthermore, private organic organizations are permitted to affix their own labels in addition to the USDA label, indicating that the product meets their standards as well as the national ones. The private label may indicate only that the organization's standards are in addition to (but not superior to) the national standards. AMS reviews certification agents for re-accreditation every five years. AMS enforces the regulations by revoking or suspending a producer's certification or an agent's accreditation if a satisfactory solution to a program violation cannot be found.

The NOP final rule became effective on February 21, 2001; the program itself became fully operational on October 21, 2002. In the first step toward implementation, USDA accredited private and state certification agents, who in turn began to certify organic producers and handlers according to the standards found in 7 CFR 205. After October 21, 2002, all products sold as organic had to be in compliance with the regulations and carry the "USDA Organic" seal.

Congressional Action on Related Legislation

Cost-Sharing Start-Up Costs. Although the OFPA requires the cost of the National Organic Program to be fully supported by user fees collected for USDA accreditation and certification services, Congress has appropriated funds on several occasions to help the program in its initial stages. The FY2001 USDA appropriations act (P.L. 106-387) contained $639,000 to cover accreditation costs. In FY2002, under the Agricultural Management Assistance Program authorized by the Federal Crop Insurance Act (P.L. 106-224), Congress made $1 million available to state agriculture departments in 15 designated states to help defray the costs of certification for small-scale producers and processors.

The 2002 farm bill (P.L. 107-171, the Farm Security and Rural Investment Act), which was enacted in May 2002, also provided additional funds to support program start-up. Title X of the farm act gave USDA authority to continue to defray the costs of producers and handlers seeking organic certification through FY2007, and authorized a one-time, mandatory transfer of $5 million from the Commodity Credit Corporation (CCC) to establish a national organic certification cost-share program under the NOP. Federal funds may not cover more than 75% ($500 maximum) of a producer's or handler's costs for becoming certified. The transfer occurred in FY2002 and remains available until fully expended, which is expected to occur in 2006.[11]

Value-Added Producer Grants. The rural development title of the 2002 farm bill established a competitive grant program to promote research on the development and marketing of value-added agricultural products, to be funded through an annual transfer of $40 million from the CCC to USDA through FY2006. Projects on organically produced commodities are eligible for these grants. Congress authorized $40 million in mandatory CCC funds to be made available each year through FY2007, but the actual funding levels so far have been volatile. The program received $50 million in mandatory funds in FY2003, consisting of $40 million in unspent funds from the first CCC transfer in FY2002 and $10 million of the $40 million CCC transfer for FY2003. In FY2004 and FY2005, appropriators prohibited the spending of the mandatory funds and instead appropriated about $15 million annually.

Research. The research title of the 2002 farm bill renewed expiring authority for a competitive grant program to support research and extension activities on organic production, processing, and international marketing. The conference report also added language calling for an emphasis on classical and advanced research on genetics to improve organic crops; research to identify the marketing and policy constraints on the organic industry; and expanded on-farm research. The act authorized $3 million to be transferred annually from the U.S. Treasury to USDA, beginning in FY2003, to support this research.[12] Other provisions in the research title: (1) require USDA's Economic Research Service (ERS) to gather and maintain segregated data on the production and marketing of organic agriculture; and (2) require ERS and the National Agricultural Library to make it easier for U.S. organic producers, researchers and extension professionals to obtain the results of organic research conducted in foreign countries.[13]

Concerning the availability of production and marketing data on organic agriculture, an official of the Organic Trade Association (OTA) states that many gaps exist, and that the Economic Research Service still is relying extensively on external data sources for much of its information. As of September 2005, a comprehensive USDA survey of the entire organic sector has not been conducted. USDA's Agricultural Marketing Service has not yet begun to provide market news on the organic sector, which would permit up-to-date price discovery. Separate export and import data for organic products are not being collected at the borders. These deficiencies hamper business planning and expansion, complicate crop insurance premium-setting and loss payments, among other issues, according to OTA.

Exemption from Check-Off Programs. The 2002 farm bill contained a provision concerning issues related to the organic industry and USDA commodity research and promotion programs. These are programs that support generic advertising to promote an agricultural product (e.g., the milk mustache ads). They are funded by assessments that producers, processors, other handlers, and frequently importers, are required to deduct from revenue at the time of sale (thus they usually are called "check-off" programs). Congress has passed many laws authorizing national check-off programs for various farm commodities; there currently are 15 in operation. (For more information on check-off programs, see CRS Report 95-353, *Federal Farm Promotion ("Check-off") Programs*, by Geoffrey S. Becker.)

The Senate version of the 2002 farm bill (S. 1731) would have established a check-off program for organic commodities, but the provision was not adopted in conference. The enacted bill instead included language exempting producers and handlers who have a certified 100% organic operation from having to pay assessments under any existing commodity check-off programs in which they currently participate. For example, a producer who grows peaches on an entirely organic farm that has been certified under the NOP would be eligible to be exempted from paying an assessment under the marketing order for peaches and nectarines, when he sells his crop to a wholesaler or other handler. The proposed rule on the exemption for producers appeared in the *Federal Register* in April 2004 (69 FR 22690). Final rules for both producers and handlers of organic commodities were published in January 2005 and became effective in February 2005 (70 FR 2744 and 2763, respectively).

Livestock Feed Regulations Changed, Then Reversed. The conference report accompanying the Consolidated Appropriations Resolution for FY2003

(P.L. 108-7/H.J.Res. 2; H.Rept. 108-10; February 21, 2003), contained a provision that made a controversial change in the National Organic Program. Section 771 of the law effectively deleted the NOP standard that requires livestock and dairy producers to provide 100% organically produced feed for their animals in order to qualify for organic certification. It further provided that, if USDA surveys found that in some geographic areas organic feedstuffs cost more than twice what conventionally produced feed cost, then livestock and dairy producers in that area could be certified as organic without having to meet the NOP standard.

Reaction to this provision from several Members of Congress, the organic industry, and a few large food processors who had already committed resources to complying with the original rule, was swift and negative. Senator Leahy, who championed OFPA and its inclusion in the Food, Agriculture, Conservation, and Trade Act of 1990, circulated a "Dear Colleague" letter on February 12, 2003, urging Senators to remove Section 771 from H.J.Res 2, but without success. On February 26, 2003, Senator Leahy introduced S. 457, the Organic Restoration Act of 2003, and on February 27, Representative Farr introduced H.R. 955, a bill to repeal Section 771. The conference report accompanying emergency wartime supplemental appropriations for FY2003, which the President enacted on April 16, 2003, contained the language from H.R. 955 repealing the earlier action taken in P.L. 108-7 (Section 2104 of H.Rept. 108-76; P.L. 108-11).

USDA released the report on organic feed availability in June 2003. It states that "ample acreage is available to provide more than enough feed grains to meet the needs of organic livestock and broiler producers ... [and] with limited exceptions, prices for organic poultry rations are not more than twice the prices of convention poultry feed rations."[14]

USDA Regulatory Activity

Interpretive Issue Statements. In April 2004, the National Organic Program headquarters in USDA released "issue statements," i.e., guidelines for interpretation of program regulations in four areas. These were livestock feed, livestock health, inert ingredients in approved organic pesticides, and the inclusion of non-food items like lotions and cosmetics within the scope of the NOP. AMS issued the statements administratively (rather than through the *Federal Register*). Many program participants were immediately critical of the Department for not giving the issue statements to the National Organic Standards Board for advance review, since the critics viewed them as revisions of the regulations, not simply as clarifications. USDA rescinded the issue statements in

May 2004, and asked the Board to provide feedback on them, which it did in October 2004.

Subsequently, on August 23, 2005, the Department issued a memorandum reversing its original position on the "scope of program" issue.[15] The memorandum states that, if all standards for organic agricultural ingredients are met, then the organic seal can be used on personal care products, supplements, and pet foods. This reversal was prompted by a lawsuit that was brought against USDA after it released the issue statement in April 2004 saying that nonfood organic products could not carry the "USDA Organic" seal.

The sequence of events is at issue here, more so than the issue statements themselves. The NOP is a new program covering a sector of U.S. agriculture that wants to be perceived as decidedly different from conventional agriculture, which has been covered by USDA marketing and regulatory programs for almost a century. Unlike other marketing programs, the NOP regulations are much more prescriptive regarding production and manufacturing practices, and NOP officials understandably might still be learning how to deal with these new factors in a regulatory setting. Thus it might be expected that NOP officials could on occasion take administrative action without thinking it necessary to consult the National Organic Standards Board, or might differ with the Board on the form and substance of a proposed regulation. Situations of this sort were largely responsible for the Department's taking 11 years to publish an NOP final rule after OFPA enactment in 1990. Similarly, the National Organic Standards Board and industry stakeholders might perceive program officials' failure to consult the Board on how certain regulations should be interpreted as dismissive of their input.

An audit report on the NOP released in July 2005 by USDA's Office of Inspector General (OIG) directly addresses these concerns.[16] The report found that "AMS has not established protocols for working with the National Organic Standards Board or resolving conflicts with them," and recommends that the NOP "establish procedures for receiving, reviewing, and implementing recommendations from the Board." The report states that "AMS also needs to improve management controls for administering the NOP," and recommends that AMS "resolve and implement internal operating procedures for such things as the resolution of complaints to govern program operations." The OIG made 10 recommendations in all. AMS officials agreed with them and promised to implement solutions by the end of FY2005, or publish draft procedures for public comment by the end of December 2005.

Peer Review of Accreditation. Section 2117of OFPA (7 CFR 205.509) requires that a three-person panel periodically review NOP's accreditation procedures, site evaluation reports, and decisions, in order to determine their adherence to the regulations. This part of OFPA has not been implemented to date because the statutory language is vague concerning some basic issues; e.g., how to find a sufficient number of experts in organic accreditation who themselves are not USDA-accredited certification agents.

In lieu of the three-person panels, USDA contracted in 2003 with the American National Standards Institute (ANSI), a private, non-profit organization that administers the U.S. voluntary standardization and conformity assessment system (part of a world-wide system commonly referred to as ISO 9000; ISO stands for International Standards Organization).[17] ANSI conducted a peer evaluation of the NOP accreditation system and delivered its report in December 2004.[18] It found a number of discrepancies between NOP's accreditation policies and procedures and those outlined in ISO 9000. Furthermore, ANSI reported that it had difficulty assessing the discrepancies because USDA has not fully documented the quality system it uses to assess applicants for accreditation.

The question of how to implement the original statutory language remains to be addressed by NOP officials and the National Organic Standards Board.

Controversy over "Access to Pasture"

In January 2005 a newspaper article about a Colorado organic dairy operation raised a major controversy within the industry and among some public interest groups.[19] The article focused on a 5,300-cow organic farm where the animals were fed almost exclusively on grain and allowed outdoors mostly for exercise.

The NOP regulation at the core of the dispute is 7 CFR 205.239(a)(1-2): "The producer of an organic livestock operation must establish and maintain livestock living conditions which accommodate the health and natural behavior of animals, including (1)Access to the outdoors, shade, shelter, exercise areas, fresh air, and direct sunlight suitable to the species, its stage of production, the climate, and the environment; (2) Access to pasture for ruminants[.]"

The news article cites organic dairy producers who argue that the regulation means that organic cattle must get some of their nutrition, as well as fresh air and sunshine, from grazing on pasture. The Colorado operator maintains that his animals have outdoor access, but that in an arid state like Colorado, providing sufficient pasturage for all his cows to graze would be an insurmountable requirement.

At a February 2005 meeting of the National Organic Standards Board, members discussed and recommended new language for 7 CFR 205.239, which it forwarded to National Organic Program officials for approval.[20] The emphasis in the revised language was on allowing cows at the appropriate "stage of life" to graze pasture "during the pasture's normal growing season." At the August 2005 Board meeting, the NOP staff rejected the recommendation saying it lacked a "clear and concise regulatory objective," and asked the Board to rework it.[21] It could be a year before a proposed rule reflecting a revised NOSB recommendation could be published for comment in the *Federal Register*, a delay that some speakers at the Board meeting argued was unacceptable. A guidance document that the Board posted on its website for comment in March 2005 is serving unofficially as an interim interpretation of the current "access to pasture" requirement until a proposed revised standard can be published.[22]

The issue is sensitive because it involves the question of farm size. USDA considers the NOP to be a marketing program that is size-neutral. On the other hand, many in the industry hold that organic farming *practices* are an integral part of the meaning of the term "organic," particularly with respect to standards for the treatment and feeding of livestock. Some observers argue that if the pasturage in certain parts of the nation can support only small dairy or beef cattle herds, or none at all, then such farms should be small or nonexistent in those areas. Others are concerned lest the regulations become so prescriptive that they deprive producers and processors of the opportunity to benefit from the expanding market for organic products.

COURT RULING CONTROVERSY

In June 2005, a First Circuit Court sided with an organic farmer who had filed suit against USDA claiming that several provisions of the final NOP rule were more lenient than the underlying statutory language allowed.[23] Specifically, the court determined that certain "natural" substances not commercially available in organic form had to be individually reviewed to determine their status for the National List of Approved and Prohibited Substances before they could be used in organic-labeled products. Second, the court determined that synthetic substances that heretofore had been approved in the regulations and used in commerce could not be used in the processing or handling of organic-labeled products. And third, it ruled that the USDA regulation on converting a dairy herd from conventional to organic production did not in any way reflect the statutory language and was thus invalid. The District Court in this case set a one-year time frame for USDA to develop new regulations and allowed another year beyond that (until June 2007) for the industry to come into compliance.

Stark differences of opinion within the organic industry on the impact of the court judgement became immediately apparent. Representatives of consumer groups and some food retailing groups welcomed the decision. They maintained that consumers consider the organic label to mean the absence of synthetic ingredients and that the new, stricter regulations would reconfirm consumer confidence in the OFPA as enacted. These groups have expressed strong opposition to resolving the court decision issue by amending the OFPA.

Conversely, a large number of organic food manufacturers were apprehensive that the ultimate outcome of the court decision would be to force the discontinuation of hundreds of existing, organic-labeled food and beverage products, and/or end manufacturers' ability to use the "USDA Organic" seal, which commands premium prices in the marketplace. These stakeholders supported the efforts of the Organic Trade Association to have Congress effect a legislative solution to the issues raised by the court decision.

In late September 2005, during floor debate on its version of the FY2006 USDA spending bill (H.R. 2744), the Senate adopted an amendment requiring the Secretary to conduct a survey to determine the impact of the decision and of various approaches to addressing it, including amending the OFPA to effectively undo the court's action.

In late October 2005, conferees on the FY2006 agriculture appropriations bill adopted a second provision in committee. The overall effect of this provision is to change the statutory language to make moot the latter two of the three court holdings. Specifically, the provision amends the OFPA: (1) to allow "natural" but non-organic products to be used as long as they are subject to the National List evaluation process; and additionally, to permit USDA to develop an expedited procedure for approving the addition of such products to the National List for a limited time period; (2) to remove the original language that generally prohibits the inclusion of synthetic substances on the National List (although there are other requirements that must be met); and (3) to provide a statutory basis for a new regulation on converting dairy herds to organic that is related to on-farm production of organic feed and forage.

CONCLUSION

It will be necessary for USDA to write new regulations to address these three areas. Barring further legislative changes, the new rules will be based on revised statutory language. The history of the NOP, however, indicates the difficulty in predicting the outcome of the notice and comment process. This area of public policy may remain contentious, depending upon the ability of the organic community's diverse stakeholders to develop a workable consensus among themselves on interpreting the OFPA. Additional debate on the statute could happen when Congress begins consideration of the next farm bill, expected in 2007. Hearings on various parts of a farm bill could begin in 2006.

NATIONAL ORGANIC PROGRAM FAQS

The questions and answers below are categorized by subject matter and will be updated on a monthly basis.

General Topics

Q: Where can I get a copy of the National Organic Program regulations?

A: You can obtain a copy of the NOP regulations by contacting the NOP through this website.

Definitions

There are no Questions and Answers for Definitions at this time

Applicability

Posted 6/26/03

Q: Can salt and water be certified organic under the National Organic Program?

A: No. Salt, in all of its forms, and water can not be certified as organic under the National Organic Program. Only agricultural products produced and handled in accordance with the National Organic Program may be certified as organically produced. Salts are minerals not agricultural products. Water is a chemical substance used in the production of agricultural products but is not itself an agricultural product. Section 2106(c) (7 U.S.C. 6505 (c)) of the Organic Foods Production Act of 1990, as amended, and section 205.302 of the National Organic Program require the exclusion of salt and water when calculating the percentage of organic ingredients in a multi-ingredient product. Further, National Organic Program sections 205.303(b)(1) and

205.304(b)(1) provide that “Water or salt included as ingredients cannot be identified as organic.”

Q: When must organic producers and handlers be certified to continue to market their product as organic?

A: Beginning on October 21, 2002, producers and handlers must be certified by a USDA-accredited certifying agent to sell, label, or represent their products as "100 percent organic," "organic," or "made with organic (specified ingredients or food group(s)." Please see section 205.101 for exemptions from certification.

Q: I am a small farmer. Will I have to be certified?

A: It depends. If your gross agricultural income from organic sales total $5,000 or less annually, you are exempt from certification (see section 205.101(a)(1) of the NOP regulations). Exempt operations must comply with the applicable requirements of subpart C and the labeling requirements at section 205.310 of the NOP regulations.

Q: I am a small processor. Do I have to be certified?

A: It depends. If your gross agricultural income from organic sales total $5,000 or less annually, you are exempt from certification (see section 205.101(a)(1) of the NOP regulations). Exempt operations must comply with the applicable requirements of subpart C and the labeling requirements at section 205.310 of the NOP regulations. For other possible handler exemptions see section 205.101 of the NOP regulations.

Q: I sell less than $5,000 worth of organic product. Can I be certified?

A: Yes. Any qualified agricultural production or handling operation may be certified as an organic production and handling operation. This includes all qualified production and handling operations eligible for exemption or exclusion under section 205.101 of the NOP regulations.

Q: Can non-certified companies use the word "organic?"

A: Producers and handlers that qualify for exemption or exclusion from certification may use the term "organic" in compliance with the labeling requirements specific to their exemption or exclusion (see section 205.101 of NOP regulations).

Q: What type of records would be acceptable to National Organic Program as proof of exempt organic operations showing compliance with the labeling requirements, production requirements, handling requirements, record keeping, audit trail, and non use of prohibited substances.

A: Examples of records to be kept by exempt operations are listed in the preamble of Subpart B, Applicability. Exempt operations must comply with the applicable organic production and handling requirements of subpart C of the national standards and meet the labeling requirements of section 205.310. Therefore, each exempt operation should maintain records which demonstrate compliance with the Organic Foods Production Act (OFPA) of 1990 and the national standards. It is the decision of the exempt operation as to which records it needs to demonstrate compliance.

Q: Are exempt operations subject to National Organic Program (NOP) audit? If so, what fee would be charged to those operations?

A: Yes. Exempt operations that produce or handle agricultural products to be sold, labeled, or represented as "100 percent organic" or "organic" are subject to NOP compliance audits. Costs associated with compliance audits would be borne by the NOP.

Q: How was the gross receipt threshold of $5,000 established? What steps would be necessary to raise the limit?

A: The $5,000 producer or handler exemption in §205.101(a) of the national standards was mandated by OFPA (7 U.S.C. 6505(d)). Actions necessary to raise the limit of the $5,000 producer/handler exemption would include congressional amendment of OFPA §6505(d).

Q: Please explain who may use the term organic and how the term is to be used.

A: Any production or handling operation certified according to the provisions of subpart E, Certification, may use the term "organic" (§205.100). Production or handling operations that are exempted or excluded under §205.101 may use the term "organic" according to the regulations specified in §205.310, Labeling; provided, they comply with the production and handling requirements of subpart C of the national standards.

Q: What are the penalties for misuse of the term "organic"?

A: Any operation that knowingly sells or labels a product as "organic", except in accordance with the Act (OFPA) and the national standards, may be subject to a civil penalty of not more than $10,000 per violation and the provisions of 18 U.S.C 1001.

Q: Who will be responsible for the enforcement of the National Organic Program and how will a typical prosecution proceed?

A: USDA, accredited certifying agents, and where applicable, approved State Organic Programs will be responsible for enforcement of the national regulations. Compliance procedures for certified organic operations, accredited certifying agents, and State Organic Programs are specified in sections 205.660 through 205.668 of the national standards.

Q: When a retail establishment markets products supplied from a certified producer, in the event the producer is found non-compliant, will the retailer be subject to any legal recourse from the NOP? Do the same rules apply (to the retailer) for products produced by exempt and non-exempt producers?

A: If a provider of product to a retail food establishment is found to be in violation of the national organic standards and the retail food establishment is not a party to that violation, there will be no action by the National Organic Program (NOP) against the retail food establishment. This holds true for certified producers and handlers as well as those claiming exemption under section 205.101.

Any person, including a retail food establishment, who knowingly sells or labels a product as organic, except in accordance with the OFPA and the national organic standards, shall be subject to a civil penalty of not more than $10,000 per violation.

Products that have entered the channels of commerce before the certified operation's suspension or revocation will not result in a product recall, unless the non-compliance involves a food safety issue. For further information see page 80627 of the national organic standards.

Organic Production and Handling Requirements

Posted 11/29/05

Q: With Avian Influenza becoming an increasingly serious issue in the poultry industry, what can organic poultry producers and their certifying agents do to help recognize and prevent the spread of this disease?

A: USDA's Animal and Plant Health Inspection Service has provided guidance to poultry producers, including free range and organic producers, regarding biosecurity considerations for their operations – this guidance is found at http://www.aphis.usda.gov/vs/birdbiosecurity/

Integral to the National Organic Program is a management plan for each organic production facility – among other things, this plan must address preventative livestock health care practices, including sanitation practices that will be followed to minimize the occurrence and spread of diseases and parasites.

USDA has brought the APHIS biosecurity guidance materials to the attention of its accredited organic certifying agents for their use in working with organic poultry producers.

If it is determined by a producer and certifier that for the health, safety, and welfare of poultry there is need for temporary confinement of birds, provision exists within the National Organic Program for such confinement without a loss of organic certification.

If Federal or State animal health authorities determine that more prescriptive actions are required, organic producers could be required to adhere to those actions.

Posted 1/14/05

Q: If a producer adheres to all aspects of the National Organic Program (NOP), including never utilizing biotech-derived seeds, but a certifying agent tests and detects the presence of biotech-derived material in the crop, is that crop's status determined to be no longer "certified organic?" And, if so, what in the NOP supports this conclusion?

A: It is particularly important to remember that organic standards are process based. Certifying agents attest to the ability of organic operations to follow a set of production standards and practices that meet the requirements of the Act and the regulations. This regulation prohibits the use of excluded methods in organic operations (§205.2—Terms defined, and §205.105—Allowed and prohibited substances, methods, and ingredients in organic production and handling). The presence of a detectable residue of a product of excluded methods alone does not necessarily constitute a violation of this regulation. As long as an organic operation has not used excluded methods and takes reasonable steps to avoid contact with the products of excluded methods as detailed in their approved organic system plan, the unintentional presence of the products of excluded methods will not affect the status of the organic operation. As to the status of the commodity, USDA's position is that this is left to the buyer and seller to resolve in the marketplace through their contractual relationship. (See page 80556 of the preamble, "Applicability—Clarifications; (1) "Genetic drift").

Posted 1/14/05

Q: Do insufficient buffers or barriers that result in unintended contact with a product of genetic modification threaten the farm's certification or use of the field for the production of organic crops? If an organic producer or handler is found to have not implemented measures necessary to prevent commingling of organic and non-organic products, would that threaten the certification of the producer or handler?

A: In order to become a certified organic operation, a producer must submit an Organic System Plan (plan) to a USDA-accredited certifying agent for approval. That plan must include, among other things, evidence that sufficient buffer zones have been incorporated into the operation to ensure the integrity of the organic crop operation. The certifying agent must not approve a plan that does not provide evidence of sound measures taken to ensure the integrity of the organic crop operation, including buffer zones and other steps to prevent commingling with unapproved non-organic materials or conventional crops. If a producer does not adhere to such preventive measures, the certifying agent is expected to denote such failure as a noncompliance and take appropriate measures toward correction by the producer. Inadequate buffer zones should not be approved in the first place and failure to comply with approved buffer zones constitutes a noncompliance with the approved organic system plan. (See the preamble, page 80558, on Subpart C—General Requirements, which describe what must be contained in an organic system plan, and §205.2 under terms defined – Buffer zone.)

However, even when all precautions have been taken, and an approved buffer zone fails to provide the protection that both the operator and the certifying agent reasonably expected, certifying agents must not "retroactively" punish the producer by an enforcement action or "de-certify" the organic crop. The appropriate action to take in this case is to re-evaluate the buffer zone and other preventive measures in the plan to ensure improved integrity and performance in the future. As to the status of the commodity, USDA's position is that this is left to the buyer and seller to resolve in the marketplace through their contractual relationship. (See page 80556 of the preamble, "Applicability—Clarifications; (1) "Genetic drift").

Posted 1/14/05

Q: Has any certified organic operation that refrains from intentional use of biotech seeds ever lost certification for the inadvertent presence of biotech material in its crop?

A: No accredited certifying agent has reported to us that certification has been lost due to adventitious presence of biotech material. In one instance, a producer admitted to deliberately planting GM seed and representing the crop as organic, for which we took enforcement action and he is no longer certified.

Posted 1/14/05

Q: Are food labels stating "GM, GE, or GMO-free" part of the National Organic Standards?

A: They are not. Truthful labeling is embodied in the National Organic Standards, as supported by USDA's Food Safety and Inspection Service (FSIS), the Food and Drug Administration (FDA), and the Federal Trade Commission (FTC) – the agencies with respective jurisdiction over truthful labeling laws. In the preamble of the National Organic Program final regulations, we stated that organic is not synonymous with "GM-free," when we said: "These phrases may…be used as additional, eco-labels, provided they are truthful statements…[but] they are not permitted as replacements for the term 'organic.'" (See page 80586 of the preamble, under "Labeling—Changes Requested But Not Made: (7) Use of Other Terms as Synonymous for "organic").

Posted 1/14/05

Q: Is there a working definition of the word "contamination" within the NOP? Are all products of genetic modification considered "prohibited substances" as defined in the final regulations? And, what actions are authorized or required when organic crops or products are found to contain unintended or inadvertent genetically modified hybrids or other genetically modified substances?

A: There is no definition in the final regulations of the National Organic Standards for the word "contamination." By our count, "contamination" is mentioned nearly 50 times in the regulations.

All genetically modified practices or products are indeed considered prohibited, as cited in 205.105, the paragraph that describes "excluded methods." Please refer back to the above issue when considering the adventitious presence of a genetically modified or genetically-engineered substance. Such adventitious presence does not affect the status of the certified operation and does not necessarily result in loss of organic status for the organic product, provided it was produced in adherence with all of the organic requirements under 7 CFR 205.

Again, the action regarding the final product's status in this case is left to the determination by the buyer and seller of the product. Contamination by a prohibited substance, when mandated by a government body, however, would result in loss of organic status for the product, even when all other regulations had been followed. In the case of an emergency spray program, for example, if the spray is a prohibited substance but is mandated by a State or Federal program, the crop's organic status is lost and that crop must be diverted for sale in the conventional market. Neither the operation nor the land's organic status is altered by an emergency spray program, however. (See §205.672 Emergency pest or disease treatment.)

Posted 3/6/03

Q: Do I have to subtract the use of steam condensate when calculating the percentage of organically produced ingredients? For example; blanching fruits or vegetables, cooking organic ingredients with steam, or other uses for processing with steam.

A: No.

Posted 1/6/03

Q: One of my clients extracts the oil from organic mint leaves. Although he suspected that a recent crop of mint leaves may have come in contact with a prohibited pesticide, residue testing showed that the unprocessed mint leaves had a residue level which does not exceed 5 percent of the EPA tolerance for that particular pesticide.

Is it possible that although the pesticide residue level was okay in the raw mint leaves, the mint oil could contain a residue level above 5 percent of the EPA tolerance level, meaning the mint oil couldn't be sold as organic?

A: It is possible. The Environmental Protection Agency has established pesticide residue tolerance levels for many processed products (mint oil is a processed product), and these tolerance levels are different from the levels established for the raw product from which the processed product is made.

In this case, there is a separate tolerance level established for mint oil, and the mint oil cannot contain pesticide residue greater than 5 percent of the EPA tolerance established for mint oil for that particular pesticide and still be sold as organic.

When there is no EPA tolerance level established for a processed product, the tolerance level established for the raw product from which the processed product is made will be used to determine if the pesticide residue level is acceptable for that processed product.

EPA pesticide residue tolerance levels can be found at www.access.gpo.gov/nara/cfr/waisidx_01/40cfr180_01.html.

Posted 12/9/02

Q: I operate a certified organic farm, and I wish to purchase compost from a commercial compost operation. Does this compost have to be certified organic?

A: No. Although a commercial compost operation may become USDA certified, there is no requirement for it to do so, and there is no requirement for you to use certified organic compost.

However, any compost you use must meet all the requirements of the NOP regulations, section 205.203. The burden of proving that a compost used on a certified organic operation meets all the requirements of section 205.203 rests with the certified organic operation that uses the compost. That operation must show in its organic systems plan, and be able to prove to its certifying agent, that the compost it uses meets the NOP requirements.

Posted 10/18/02

Q: Must seeds for cover crops be organic?

A: Yes, unless conditions meet those outlined in Section 205.204, provided below:

§ 205.204 Seeds and planting stock practice standard.

(a) The producer must use organically grown seeds, annual seedlings, and planting stock: Except, That,

(1) Nonorganically produced, untreated seeds and planting stock may be used to produce an organic crop when an equivalent organically produced variety is not commercially available, Except, That, organically produced seed must be used for the production of edible sprouts;

(2) Nonorganically produced seeds and planting stock that have been treated with a substance included on the National List of synthetic substances

allowed for use in organic crop production may be used to produce an organic crop when an equivalent organically produced or untreated variety is not commercially available;

(3) Nonorganically produced annual seedlings may be used to produce an organic crop when a temporary variance has been granted in accordance with § 205.290(a)(2);

(4) Nonorganically produced planting stock to be used to produce a perennial crop may be sold, labeled, or represented as organically produced only after the planting stock has been maintained under a system of organic management for a period of no less than 1 year; and

(5) Seeds, annual seedlings, and planting stock treated with prohibited substances may be used to produce an organic crop when the application of the materials is a requirement of Federal or State phytosanitary regulations.

Posted 10/16/02

Q: Can wood treated with fungicides or other prohibited substances be used for fence posts, trellis systems, etc., in organic production?

A: As provided in Section 205.206(f), the producer must not use lumber treated with arsenate or other prohibited materials for new installations or replacement purposes in contact with soil or livestock. The Preamble (Crop Production--Changes Based on Comments (7)) elaborates on 205.206(f) as follows:

> "This provision prohibits the use of lumber treated with arsenate or other prohibited materials for new installations or replacement purposes in contact with an organic production site. We included this modification to clarify that the prohibition applies to lumber used in direct contact with organically produced and handled crops and livestock and does not include uses, such as lumber for fence posts or building materials, that are isolated from production. The prohibition applies to lumber used in crop production, such as the frames of a planting bed, and for raising livestock, such as the boards used to build a farrowing house."

Q: Do fields have to meet any size requirements in order to produce crops certified as organic?

A: No. There are no field-size requirements relative to whether or not an operation can be certified organic.

Q: Is the feed that is fed to organic livestock and poultry "vegetarian" feed?

A: Not necessarily. Although as stated in the National Organic Standards subpart C, section 205.237(b)(5), "The producer of an organic operation must not feed mammalian or poultry slaughter by-products to mammals or poultry." there is no restriction against organic livestock feed containing appropriate fish products.

Q: I know organic agriculture prohibits the use of GMOs, but do the guidelines allow for a small amount of contamination?

A: As GMO contamination of organic crops relates to genetic drift, the Preamble to the National Organic Program regulations, Applicability, Clarifications (1) Genetic Drift, states:

> "This regulation prohibits the use of excluded methods [which include GMOs] in organic operations. The presence of a detectable residue of a product of excluded methods alone does not necessarily constitute a violation of this regulation. As long as an organic operation has not used excluded methods and takes reasonable steps to avoid contact with the products of excluded methods as detailed in their approved organic system plan, the unintentional presence of the products of excluded methods should not affect the status of an organic product or operation."

However, if a certifying agent has reason to suspect that an organic product has come into contact with prohibited substances or been produced using excluded methods, the certifying agent can call for testing, which under certain conditions could result in that product no longer being considered "organic."

For a complete discussion of this issue, you must read the National Organic Standards, subpart G, Administrative, sections 205.670 – 205.671.

Q: Can reclaimed water be used for irrigation on organic farms?

A: Generally, the National Organic Standards place no further restrictions on reclaimed water used for irrigation beyond those imposed by State Departments of Natural Resources. However, to fully answer this question, we would need to know how and from what source the water is being reclaimed.

Q: We are a meat processing plant interested in processing all natural pork and beef. What kind of general information can you give us?

A: There are now specific USDA standards that must be met by all producers and processors who wish to label their agricultural products as organic. (Please note that "natural" is not synonymous with "organic.") These producers and handlers must be certified by USDA-accredited certifying agents by October 21, 2002. The organic standards for producing and processing livestock can be found in the National Organic Standards, subpart C, Sections 205.236 – 205.272.

A list of USDA accredited certifying agents can be found on the NOP website at www.ams.usda.gov/nop.

Q: Must manure used to fertilize organic crops come from an organic source, that is from an animal that has been raised in accordance with the organic standards?

A: No. There are no restrictions as to the source of manure. For greater detail on soil fertility and crop nutrient management, please read the National Organic Program regulations, subpart C, section 205.203. This topic is further discussed in the Preamble to the National Organic Program regulations, under the section headed Crop Production, Changes Requested but Not Made, (2) No Prohibition on Manure from Nonorganic Operations.

Q: Are there any known health hazards associated with using animal waste as fertilizer for organic crops?

A: Properly managed animal waste should not cause any known health hazards. Animal waste used to fertilize organic crops must be managed in accordance with the National Organic Program regulations' soil fertility and crop nutrient management practice standard, as stated in the National Organic Program regulations, subpart C, section 205.203(c), which states, "The producer must manage plant and animal materials to maintain or improve soil organic matter content in a manner that does not contribute to contamination of crops, soil, or water by plant nutrients, pathogenic organisms, heavy metals, or residues of prohibited substances." See this section for a description of plant and animal materials, including manure and compost.

Q: Is it safe to say that using organic fertilizers and other organic farming practices is better for the soil and less of a threat to ground and surface water than commercial methods?

A: USDA's National Organic Program is a marketing program and makes no claims that organic farming is "better" in any respect than conventional farming.

Q: Has the National Organic Program issued any guidance or sample plans regarding the development of production and handling system plans?

A: The National Organic Program has issued no specific guidance or samples relating to the development of organic production and handling system plans. However the NOP is helping to fund the development of crop and livestock checksheets designed for organic farmers to use in assessing their compliance with the National Organic Standards. The checksheets are being developed by the National Center for Appropriate Technology (NCAT). For more information, visit NCAT's website at www.ncat.org.

Q: Is the use of ionizing radiation allowed at any stage of production or handling of agricultural products labeled organic?

A: No. Irradiation in the production and handling of organic food is prohibited by the National Organic Standards, subpart B, Applicability, section 205.105, Allowed and prohibited substances, methods, and ingredients in organic production and handling, which states, "To be sold or labeled as '100 percent organic,' 'organic,' or 'made with organic (specified ingredients or food group(s)),' the product must be produced and handled without the use of: (f) Ionizing radiation, as described in Food and Drug Administration regulation, 21 CFR 179.26 ...

Q: Under the National Organic Standards, can the certifying agent approve ion exchange for the processing of organic foods?

A: The National Organic Standards do not prohibit the use of ion exchange in organic food processing as a technology. However, the National Organic Standards do prohibit the use of synthetic substances used in or on processed products unless included in National List section 205.605 as an allowed substance. Accordingly, ion exchange may not be used in the processing of organic foods unless the synthetic substances used in the ion exchange process are listed as approved substances in National List section 205.605.

Q: Can I use homeopathic treatments to treat my livestock?

A: The National Organic Standards do not prohibit use of homeopathic treatments in the production of organic livestock. The producer must make certain that such treatment(s) does not contain a prohibited substance. (Section 205.105(a))

Q: Can I remove organic calves from my farm, raise them conventionally for a year, bring them back to the farm, and then manage them organically for one year prior to the production of organic milk and milk products?

A: No. Livestock or edible livestock products that are removed from an organic operation and subsequently managed on a nonorganic operation may not be sold, labeled, or represented as organically produced. (Section 205.236(b)(1))

Q: Can I sell an organic dairy animal as slaughter stock?

A: Dairy animals that have been under continuous organic management since the last third of gestation may be sold, labeled, or represented as organic slaughter stock. Conversely, dairy animals that have not been under continuous organic management since the last third of gestation may not be sold, labeled, or represented as organic slaughter stock. (Section 205.236(b)(2))

Q: Do all dairy animals have to be organic from the last third of gestation to produce organic milk or milk products?

A: No. Milk or milk products must be from animals that have been under continuous organic management beginning no later than 1 year prior to the production of the milk or milk products that are to be sold, labeled, or represented as organic. Except that, in the case of the conversion of an entire distinct herd, the animals must receive a minimum of 80 percent feed that is either organic or raised from land included in the organic system plan and managed in compliance with organic crop requirements during the first 9 months of the 1-year conversion period. (Section 205.236(a)(2))

Q: Does Breeder Seed need to be raised organically in order for its progeny; i.e., Foundation, Registered, and/or Certified1 Seed, to be sold, labeled, or represented as organic seed?

A: No. The seed of any generation planted with conventional, untreated seed and produced under organic conditions can be certified as organic. In other words, untreated seed of Foundation, Registered, or Certified generations may be sown in an organic seed field/cage/greenhouse, and the progeny will qualify as organic seed. Many varieties are only legal for sale as a class of certified seed according to the Plant Variety Protection Act (PVP), Title V. Check with your official state seed certification agency for the specific varieties protected by PVP, Title V. Additionally, it is not necessary to use only genetically1 certified 2 untreated seed to produce organic seed. Any generation of "common," untreated seed may also be used.

1. The use of the term “genetically” in this sentence in no way implies or authorizes the use of genetically modified organisms.
2. Certified – As defined by the Federal Seed Act Regulations (7 C.F.R. 201(2) (ee)).

Q: Does organic seed always need to be used to produce an organic crop?

A: For edible sprouts, yes; for all other organic crops, no. The National Organic Standards (7 C.F.R. 205.204 (a) (1)) provides that nonorganically produced, untreated seeds may be used to produce an organic crop when an equivalent organically produced variety is not commercially available. The National Organic Standards (7 C.F.R. 205.204 (a) (2)) also provides that nonorganically produced, treated seeds may be used to produce an organic crop when an equivalent organically produced or untreated variety is not commercially available. The seed treatment, however, must be with a substance included on the National List of synthetic substances allowed for use in organic crop production. The only time that a seed treated with a prohibited substance may be used to produce an organic crop is when use of the prohibited substance is a requirement of Federal or State phytosanitary regulations.

Q: How does the National Organic Program (NOP) interpret “equivalent variety” in 7 C.F.R. 205.504(a) (1-2)?

A: An equivalent variety means a variety exhibiting the same “type” (such as head lettuce types, leaf lettuce types, etc.) and similar agronomic characteristics such as insect and disease resistance when compared to the original varietal choice.

“Type” is defined by the Federal Seed Act of 1939 (7 U.S.C. 1551-1661.) as either (A) a group of varieties so nearly similar that the individual varieties cannot be clearly differentiated except under special conditions, or (B) when used with a variety name means seed of the variety named which may be mixed with seed of other varieties of the same kind and similar character, the manner of and the circumstances connected with the use of the designation to be governed by the rules and regulations prescribed under section 1592 of the Federal Seed Act.

Variety is defined by the Federal Seed Act as a subdivision of a kind which is characterized by growth, plant, fruit, seed, or other characters by which it can be differentiated from other sorts of the same kind, for example, Marquis wheat, Flat Dutch cabbage, Manchu soybeans, Oxheart carrot, and so forth. Kind means one or more related species or subspecies which singly or collectively is known by one common name, for example, soybean, flax, carrot, radish, cabbage, cauliflower, and so forth.

Q: How many generations removed from the present must a grower research the breeding techniques for a given variety to ensure compliance with the NOP? In many cases it is impossible to research breeding techniques of older varieties as breeding methods are usually highly confidential and in some cases lost to history.

A: Effective October 21, 2002, plant breeders developing new varieties for use in certified organic production must comply with the requirements of the NOP.

All varieties (open pollinated and hybrid) in existence prior to October 21, 2002 may be used by organic producers provided that these varieties have not been produced using Genetically Modified Organisms (GMO).

What constitutes GMO is defined by NOP through the term "excluded methods," (7 C.F.R. 205.2).

Q: How detailed does my Organic Systems Plan (OSP) have to be?

A: The OSP must be sufficient in detail for the certifying agent to determine that your production or handling operation is in compliance with the applicable NOP regulations. The extent of this detail will be determined in collaboration with your certifying agent.

Q: Do I have to follow the Organic System Plan (OSP) I have filed with my certifying agent?

A: Yes. Your OSP is a detailed description of how your operation will achieve, document, and sustain compliance with all applicable provisions of the NOP regulations. Before granting certification to your operation, your certifying agent must concur that your OSP fulfills the requirements of NOP regulations. Your OSP must be annually updated and approved by your certifying agent (see section 205.406 of the NOP regulations). In addition, your OSP may be modified at any time upon request to and written approval from your certifying agent. If you deviate from your previously approved OSP without written approval from your certifying agent, you are no longer in compliance with NOP regulations and could be subject to suspension or revocation of your certification.

Q: Do NOP regulations require the use of organic seed?

A: NOP regulations require the use of organic seed when commercially available. For your options when organic seed is not commercially available see section 205.204 of the NOP regulations or consult your certifying agent.

Q: During the implementation period can a processor (handler) continue to use product ingredients sourced from production operations that have not been certified by a USDA accredited certifier, but have been certified by other certifiers?

A: Yes. During the implementation period, a handler would be allowed to source organic agricultural product ingredients from operations that have not been certified by a USDA accredited certifier. All organic handling operations must discontinue such practices and comply with NOP regulations by October 21, 2002.

Q: How will I find out about changes regarding laws, regulations, policies, and procedures?

A: The NOP will notify accredited certifying agents of changes and pending changes to laws, regulations, policies, and procedures and post such changes and pending changes to its website. Certifying agents should, in turn, notify their clients of such changes and pending changes. Amendments to the NOP regulations will require rulemaking with an opportunity for public comment. This rulemaking is published in the Federal Register. The NOP will issue news releases on Federal Register publications and post such news releases and Federal Register documents to its website.

Q: How will a National Organic Program audit of a producer be conducted?

A: Producer audits/evaluations will be conducted by accredited certifying agents according to the provisions specified in subpart E, Certification, of the national standards.

Q: When are the greenhouse production and apicultural standards expected to be published?

A: Proposed greenhouse and mushroom standards were announced by the National Organic Standards Board's (NOSB) Crops Committee at the June 2001 NOSB meeting. The proposals were posted at the NOSB link on the National Organic Program (NOP) website for public comment through July 31, 2001. The NOSB created an apiculture task force at its June 2001 meeting. Final recommendations for greenhouse, mushroom and apiculture standards will be submitted to the full NOSB for review and approval at its October 15 - 17, 2001, meeting in Washington, D.C. Pending the NOSB decision on the proposed recommendations, the NOP will begin rulemaking for greenhouse, mushroom and apiculture standards in October 2001. NOP's goal will be implementation of these standards by October 21, 2002.

Q: Please provide discussion, examples, and guidance. on buffer zones. Section 205.202 requires distinct boundaries and "adequate" buffer zones, but minimums are not specified, nor is the term adequate defined.

A: Section 202.202(c) requires distinct, defined boundaries and buffer zones to prevent the unintended application of a prohibited substance to land under organic management.

In examining this issue, USDA concluded that imposing a specific size for buffer zones could impose unnecessary burdens on some organic producers without offering greater protection of organic fields and crops from unintended contact with prohibited substances. For example, buffer zones might not be needed for an organic farm if it were completely surrounded by wilderness or areas not in agricultural production. Accordingly, the national standards do not specify specific dimensions for buffer zones, but leaves the determination of their size to the organic producer and the certifying agent on a case-by-case basis.

It has always been the responsibility of organic operations to manage potential contact of organic products with other substances not approved for use in organic production systems. The organic system plan must outline steps that an organic operation will take to avoid drift from neighboring operations, particularly drift of synthetic chemical pesticides.

When considering drift issues, both certifying agents and producers must remember that organic standards are process based. Certifying agents attest to the ability of organic operations to follow a set of production standards and practices that meet the requirements of the Organic Foods Production Act and the national standards. The national standards prohibit the use of genetically modified organisms (defined in the standards as excluded methods) in organic operations. The presence of a detectable residue of a product of excluded methods alone does not necessarily constitute a violation of the regulations. As long as an organic operation has not used excluded methods and takes reasonable steps to avoid contact with the products of excluded methods as detailed in their approved organic system plan, the unintentional presence of the products of excluded methods should not affect the status of an organic product or operation.

Therefore, while the national organic standards provide significant discretion in establishing buffer zone dimensions, buffer zones should not be sized at distances which attempt to achieve a zero tolerance for prohibited substances. The intent of the regulations are to foster a collaborative effort between the certifying agents and their grower clients to determine an appropriate buffer zone with each party being fully cognizant of the process-based nature of the organic label claim.

Q: Does the National Organic Program plan to provide any training or workshops at the producer/grower level?

A: USDA does not envision conducting any training or workshops at the producer or handler level. USDA has entered into a cooperative agreement with the National Center for Appropriate Technology to provide compliance tools for organic agricultural producers and certifying agents.

Q: How does the National Organic Program (NOP) envision that producers/growers can become familiar with the intricacies of this new regulation?

A: Under the NOP, USDA provides oversight to accredited certifying agents. Certifying agents provide oversight to producers and handlers. Section 205.501(a)(8) requires that certifying agents provide sufficient information to persons seeking certification to enable them to comply with OFPA and national organic standards. USDA has entered into a cooperative agreement with the National Center for Appropriate Technology to provide compliance tools for organic agricultural producers and certifying agents.

Q: Do nonagricultural substances included on the National List of Allowed and Prohibited Substances have to be produced without the use of volatile synthetic solvents? My certifying agent says yes because of the prohibition on the use of volatile synthetic solvents found in section 205.270(c)(2).

A: No. Section 205.270(c)(2) prohibits the use of a volatile synthetic solvent unless included on the National List as an allowed substance. However, synthetic solvents do not have to be on the National List to be allowed in the production of an allowed nonagricultural substance found on the National List. The use of volatile synthetic solvents in the production of allowed nonagricultural substances included on the National List is considered approved through the materials review process, unless otherwise stated through an annotation to the approved substance. (Example: § 205.605(a)(9), Flavors -- nonsynthetic sources only and must not be produced using synthetic solvents and carrier systems or any artificial preservative.)

Labels, Labeling, and Market Information

Posted 12/9/02

Q: We are a company that makes bread using certified organic flour and other ingredients. The yeast we use is grown and cultured using certified organic substrates, and the yeast producers follow the NOP regulations for production and processing. Using this yeast, can we label our bread 100 percent organic?

A: You do not say whether or not the yeast has been certified organic by a USDA-accredited certifying agent. If the yeast is certified organic, and all the rest of the ingredients in your bread have been certified organic, you may label your bread as 100 percent organic. If the yeast has not been certified, then your bread does not contain 100 percent organic ingredients. Therefore, you may not label your bread as 100 percent organic.

Posted 11/12/02

Q: I use flavorings in the manufacture of my organic agricultural products. I have proof that these flavorings are from nonsynthetic sources, and are produced without the use of synthetic solvents, carriers, or artificial preservatives as required in section 205.605 (a) 9. However, these flavorings do not come from a certified organic operation. In which categories of organic agricultural products am I allowed to use these flavorings (100 percent organic; organic; made with (at least 70 percent) organic ingredients)?

A: 100 percent organic—No. Products labeled 100 percent organic must contain 100 percent certified organic ingredients.

Organic—Yes. As long as the flavorings constitute 5 percent or less of the total ingredients, and meet the requirements of section 205.605(a)(9).

Made with (at least 70 percent) organic ingredients—Yes. As long as the flavorings meet the requirements of section 205.605(a)(9).

Posted 11/12/02

Q: How are producers whose company names include the term "organic" affected by the NOP regulations? For example, could a company selling organically produced vegetables call itself Blue Sky Organic?

A: The short answer is yes, as long as that company is not trying to mislead by using that name. For a more thorough explanation, I refer you to a pertinent section of the Preamble to the NOP regulations below:

Labeling - Changes Requested But Not Made

(1) "Organic" in Company Names. Many commenters stated that the term, "organic," must not be used as part of a company name if the company does not market organically produced foods. They are concerned that the term in a company name would incorrectly imply that the product, itself, is organically produced.

While we understand commenter concerns, we do not know the extent of the problem. We do not believe those concerns require such a prohibition in the regulations at this time. These regulations may not be the best mechanism to address the issue. Section 6519(b) of the Act provides the Secretary with the authority to take action against misuse of the term, "organic." USDA will monitor use of the term, "organic," in company names and will work with the FTC to take action against such misuse of the term. These determinations must be made on a case-by-case basis.

Q: On what agricultural products can the USDA Organic Seal be displayed?
A: The USDA Organic Seal may appear on organic agricultural products that are certified 100 percent organic or products that are certified as containing at least 95 percent organic ingredients.

Q: Is this seal in use now?
A: The seal may not be used until October 21, 2002.

Q: Are there any size, font, or color requirements placed on the use of the USDA Organic Seal? How can I get a copy?
A: Yes. The required specifications for the USDA Organic Seal can be found in the National Organic Program regulations, subpart D, Label, Labeling, and Market Information, section 205.311, which can be accessed at the NOP website: http://www.ams.usda.gov/nop/NOP/standards.html. A camera-ready copy of the seal is also available at http://www.ams.usda.gov/nop/Consumers/Seal.html.

Q: I know the USDA Organic Seal cannot appear on any products before October 21, 2002. Does this prohibition also apply to advertising materials?
A: The USDA organic seal cannot appear on products or on any advertising materials, including, but not limited to signage or catalogs before October 21, 2002.

Q: We want to label our poultry "natural." Where in the National Organic Standards do we look for the labeling requirements for using this term?

A: Please note that "organic" is not synonymous with "natural." There is nothing in USDA's National Organic Standards defining or regulating the use of the term "natural." USDA's Food Safety and Inspection Service (FSIS) regulates the term "natural" on meat and poultry labels.

Basically, FSIS defines "natural" in the following way: "A product containing no artificial ingredient or added color and is only minimally processed (a process which does not fundamentally alter the raw product) may be labeled natural. The label must explain the use of the term natural (such as--no added colorings or artificial ingredients; minimally processed). You will find this definition and many other labeling terms, including "free range" at the FSIS web site: www.fsis.usda.gov/oa/pubs/lablterm.htm. The FSIS labeling policy book, which goes into much greater technical detail can be accessed at: www.fsis.usda.gov.

Q: As the manager of a retail store, I purchase bulk, certified organic product, then put it in bins where consumers scoop out the amount they want. What signs can I post regarding the organic status of the product?

A: You may provide the same information as provided on the original container or shipping documents, as described in the National Organic Standards, section 205.308. For example, "If the product is prepared in a certified facility, the retail display, labeling, and display containers may use: the USDA seal; and the seal, logo, or other identifying mark of the certifying agent ..."

Q: In the produce area of my retail store, I display bulk product. I don't want to confuse consumers by displaying several different certifier names, seals and logos. What signage options do I have?

A: For non-packaged organic agricultural products, a retail store may make the same claims as those on the shipping documents or container. The USDA Organic Seal, or the certifier's logo, or both may be used for certified products. The only restriction is that the certifier's logo may not be more prominently displayed than the USDA seal (see National Organic Standards, section 205.308). As an alternative, retail stores may simply refer to a product as "organic" and not use any seals or names of certifiers.

Q: I have a deli in my retail store that makes various multi-ingredient products, where 70-95 percent of the ingredients are certified organic. These products are packaged at the customer's request, and the only thing we put on the package is price information. What organic claims can we make on the signs describing the products?

A: Your signs may indicate that the products you package are "made with (specified ingredients or food group(s))," as explained in the National Organic Standards, section 205.309. You may not list more than three ingredients/food groups. For example, you may sell chicken salad, with a sign that reads, "chicken salad made with organic chicken, celery, and grapes."

You may not represent the chicken salad as "certified organic," use the USDA Organic Seal, or the seal or name of any certifying agent (see section 205.310.)

Q: As a manager of a retail store, if I buy organic products from a small-scale organic producer who is exempt from certification, how can I label these products?

A: As explained in the National Organic Standards section 205.310, if you buy an organic product from an exempt operation, and you do not process this product any further, you may identify this product as "organic." For example, you could buy apples from an exempt producer and identify these apples to your customers as "organic." But—you may not identify these apples as being "certified organic," you may not display the USDA Organic Seal in conjunction with these apples, nor in any way represent these apples as certified organic.

If you further process these same apples, to make applesauce, for example, you may not identify the applesauce as being made with organic apples, or call it organic applesauce.

Q: Are there any restrictions on putting "place of origin" information on the label of an agricultural product that is certified organic; for example, "Organically Grown in Montana?"

A: No. The National Organic Standards do not prohibit the placement of truthful information on the labels of certified organic agricultural products.

Q. Section 205.303(b)(2) states that I must identify the certifying agent, preceded by "Certified organic by * * *" or similar phrase, on the information panel below the name of the handler or distributor. Can I meet this requirement by placing the name, acronym, logo, or seal of the certifying agent anywhere on the information panel?

A. To meet the requirements of section 205.303(b)(2), the full name or registered trade name of the certifying agent (which may include acronyms, if the acronym is a registered trade name) must be placed on the information panel below or otherwise near the information identifying the handler or distributor's name. No other printed material or information may be placed between the certifying agent's name and the name of the manufacturer or distributor. Certifying agents' logos or seals and non-registered acronyms may not be used to meet the requirements of section 205.303(b)(2).

Q. Is there a rule for rounding when calculating the percentage of organically produced ingredients?

A. As provided in section 205.302(b) of the National Organic Standards, the percentage of all organically produced ingredients in an agricultural product must be rounded down to the nearest whole number. For example, 69.99 percent would be rounded to 69 percent.

Q. To meet the requirements of section 205.303(b)(2) and 205.304(b)(2), can a handler identify the name of the certifying agent of the finished product using the following statement: "Certified organic by (insert certifying agent's name here) in accordance with the organic standards of the U.S. Department of Agriculture?"

A. Yes.

Q. With respect to application of the USDA seal on packaging material, does the term "transparent" in section 205.311(b)(2) refer to the backgrounds of the upper half of the circle and the word "organic" prior to application on packaging material, or does it refer to the transparency of the packaging material on which the USDA seal has been applied?

A. The term "transparent" in section 205.311(b)(2) refers to the backgrounds of the upper half of the circle and the word "organic" prior to application on packaging material. For example, if a certified organic handler of pasta chooses to apply the black and transparent version of the USDA seal on a brown shipping container, the brown color of the container may show through the transparent portions of the USDA seal.

Q. "As a certified handler can I make specific certification or production claims on the PDP such as Certified Organic," "Grower Certified," "Facility Certified," "Product Certified," "Organically Raised," or "Organically

Grown" in addition to or in place of "100 percent organic," "organic," or "made with organic (specified ingredients or food group(s))"?"

A. Certified operations are not prohibited from making truthful PDP label claims. However, alternative labeling claims must not be used to misrepresent a product. For example, alternative PDP label claims must not be constructed so as to mislead a consumer into thinking that a "made with" product was actually an "organic" product.

Further, the NOP may, in the future, engage in rulemaking to limit the use of alternative PDP claims when it has determined that these alternative claims are being used to misrepresent the product.

Finally, exempt or excluded producers and handlers may not represent their operations as certified or use the term "certified" when labeling their product.

Q: How do I label my products until the NOP regulations are fully implemented on October 21, 2002?

A: Until October 21, 2002, producers and handlers of organically produced agricultural products may continue their current labeling practices. All certified organic operations must discontinue the use of their current labeling practices and adhere to the labeling standards provided in subpart D, Labeling, of the Final Rule (October 20, 2002—old labeling practices allowed; October 21, 2002—old labeling practices not allowed).

Organic products that enter the chain of commerce before October 21, 2002 will not be in violation of NOP regulations. However, beginning October 21, 2002, all organic agricultural products that enter the chain of commerce must be labeled according to NOP regulations.

Q: Who should approve my label before I go to print?

A: Your certifying agent will review your label for compliance with NOP labeling requirements. Labels normally approved by a Federal agency such as FDA, FSIS, or BATF must be approved by those regulatory agencies.

Q: Can I make label claims in addition to "organic" on my product?

A: NOP regulations do not prohibit a producer or handler from making additional claims regarding their product as long as they are truthful and not misleading to the consumer. Such label claims may have to be approved by Federal agencies such as FDA, FSIS, or BATF.

Q: Do I have to include the certifier's address on the information panel of my product's package?

A: No. You must identify the certifier's name on the information panel. All other information related to your certifying agent is optional.

Q: When does organic product have to be labeled "For Export Only"?

A: A product must be labeled "For Export Only" when that product has been produced in the United States to an organic standard other than the National Organic Standards (NOS).

Q: Can domestically produced organic product that meets the NOS and an international organic standard make a claim regarding the international standard on the PDP without violating the NOS?

A: Yes. Truthful label claims are allowed under the NOS.

Certification

Posted 11/29/05

Q: With Avian Influenza becoming an increasingly serious issue in the poultry industry, what can organic poultry producers and their certifying agents do to help recognize and prevent the spread of this disease?

A: USDA's Animal and Plant Health Inspection Service has provided guidance to poultry producers, including free range and organic producers, regarding biosecurity considerations for their operations – this guidance is found at http://www.aphis.usda.gov/vs/birdbiosecurity/

Integral to the National Organic Program is a management plan for each organic production facility – among other things, this plan must address preventative livestock health care practices, including sanitation practices that will be followed to minimize the occurrence and spread of diseases and parasites.

USDA has brought the APHIS biosecurity guidance materials to the attention of its accredited organic certifying agents for their use in working with organic poultry producers.

If it is determined by a producer and certifier that for the health, safety, and welfare of poultry there is need for temporary confinement of birds, provision exists within the National Organic Program for such confinement without a loss of organic certification.

If Federal or State animal health authorities determine that more prescriptive actions are required, organic producers could be required to adhere to those actions.

Posted 3/25/03

Certification Status during the Appeals Process described in 7 C.F.R. §205.680-681

USDA's National Organic Program has received questions from producers, processors, handlers, distributors, and retailers regarding the status of a certified operation during the appeals process described in 7 C.F.R. §205.680-681.

7 C.F.R. §205.680-681 provides that persons subject to the Organic Foods Production Act of 1990 who believe they are adversely affected by a noncompliance decision of the National Organic Program's program manager, State Organic Program, or accredited certifying agent may appeal the decision to the Administrator of the Department of Agriculture's (USDA) Agricultural Marketing Service or a State Program's Governing State Official.

Certification status, including the ability to market products as organic, is not diminished during the appeals process. Certification continues in full force and effect throughout the entire process including any formal administrative proceeding conducted pursuant to USDA's Uniform Rules of Practice, 7 C.F.R. Part 1, Subpart H, or a State Organic Program's Rules of Procedure.

Posted 11/8/02

Q: I am a certifying agent who has an application from a wool producer and processor. How is wool production and processing addressed in the NOP regulations?

A: Wool is considered an inedible fiber, along with cotton, flax, etc. Inedible fibers are addressed in the Preamble to Final Rule as follows:

(6) Nonedible Fibers Products in the NOP. Some commenters asked the NOP to clarify the certification status of fibers such as cotton and flax. The final rule allows for certification of organically produced fibers such as cotton and flax. However, the processing of these fibers is not covered by the final rule. Therefore, goods that utilize organic fibers in their manufacture may only be labeled as a "made with..." product; e.g., a cotton shirt labeled "made with organic cotton."

In other words, the sheep must be certified organic in accordance with the NOP livestock standards, or the cotton, flax, etc., must be certified organic in accordance with the NOP crop standards to be identified as organic in a finished product. Also, since processing is not covered in the final rule, there are no synthetic processing aids used in fiber processing on the National List.

Posted 10/18/02

Q: I am an organic certifying agent in New England. An organic maple syrup producer has applied to me for certification. Should I require that the producer implement best management practices as recommended by the Vermont Cooperative Extension maple specialists; for example, requiring that the tree trunk reach a certain diameter before tapping, prohibiting the use of a vacuum system, or requiring that all health spouts be removed each year for proper healing of the tap-hole?

A: The National Organic Program encourages all producers to employ best management practices as outlined by the authorities in that specific field of agriculture, as long as they are not prohibited by the National Organic Program regulations. There is nothing in the NOP regulations that would prohibit any practice that you've listed in your question.

Posted 10/16/02

Q: A processing facility contracts with two farms for their production. Is it possible to certify those two farms under the name of the processing facility if they are contracted to produce only for that facility?

A: Yes. Production operations that are contracted to supply only a specific processing/handling facility can be certified as part of that unit. They must be included in the certified operation's organic systems handling plan and be inspected by the certifying agent.

Posted 10/16/02

Q: If a certified operation uses a warehouse or storage facility to store organic product must the warehouse be certified, too?

A: No. However, the warehouse/storage facility must be included in the certified operation's organic systems plan so that the certifying agent is aware of where the product is being stored and can verify that the facility is compliant with any applicable standards; for example, prevention of commingling and contact with prohibited substances, as outlined in Section 205.101 (b)(1).

Posted 10/16/02

Q: I am applying for organic certification of my farm. However, I used a substance during the 2002 growing season that had been on the OMRI list of approved substances, but is not allowed by the National Organic Program.

Can I still be certified, assuming I meet all the requirements of the National Organic Program?

A: Yes. Since the substance had previously been accepted as part of good organic farming practices and you used it in good faith, the status of your land and your eligibility for certification is not affected.

Posted 10/16/02

Q: Does a feed mill producing a feed supplement that will be mixed with feed to be fed to cows at an organic dairy need to be certified?

A: No. As discussed in the Preamble, "Livestock Production—Changes Based on Comments (4)" a natural feed additive [supplement] can be from any source, provided it is not classified as a prohibited substance on the National List, and must be in compliance with the Federal Food, Drug and Cosmetic Act.

However, if the feed mill is also producing feed to be used in organic livestock production, the mill must be certified.

Q. Can an accredited certifying agent include an expiration date or renewal date on a certificate of organic operation?

A. No. Section 205.404(c) provides that once certified, a production or handling operation's organic certification continues in effect until surrendered by the organic operation or suspended or revoked by the certifying agent, the State organic program's governing State official, or the Administrator. Accordingly, the NOP prohibits any language on the certificate of organic operation that indicates an expiration or renewal date.

In order to minimize confusion, NOP encourages all accredited certifying agents to prominently display the phrase, "Certification good until surrendered, suspended, or revoked." on the certificate of organic operation.

Q: Will organic producers and handlers apply to the USDA to be certified?

A: No. Producers and handlers will apply directly to the USDA-accredited certifying agent of their choice for certification (see section 205.401 of the NOP regulations).

Q: How long is my Organic Certificate good for?

A: Your organic certificate remains good until you voluntarily surrender your certification or your certification is suspended or revoked by the certifying agent, the State Organic Program's governing State official, or the

Administrator for violation of the Act or NOP regulations. The certifying agent will issue an updated certificate of organic operation as needed.

Q: How, and in what ways, do the national organic standards differ from the organic certification requirements for international trade, such as the European Union standards?

A: The USDA national organic standards should be considered in two segments: (1) the verification system that includes certification of organic agricultural products and accreditation of certifying agents and (2) the production, handling and labeling standards under which organic agricultural products are produced and sold. NOP is consistent with the internationally accepted guidelines for certification and accreditation, International Standardization Organization Guides 65 and 61, respectively. However, NOP requirements for production, handling, labeling and allowed and prohibited materials differ significantly from those of other countries, such as the European Union (EU), particularly in livestock production standards.

Q: If U.S. national organic standards differ from foreign organic standards, such as the EU, are there plans to standardize the regulations?

A: No. The NOP, in conjunction with the USDA Foreign Agriculture Service and the U.S. Trade Representative, has begun to establish a process through which equivalency or other trade agreements can be negotiated with governments of foreign countries to which U.S. organic products are exported.

Q: Is it possible for producers/growers to meet the U.S. national organic standards and foreign organic standards?

A: Section 205.300(b) allows for production and export of products produced in the U.S. and certified to foreign national organic standards or foreign contractor buyer requirements. Such products may be labeled in accordance with the organic labeling of the receiving country or contract buyer. Such products must be exported. The shipping containers and shipping documents must meet the labeling requirements of section 205.307(c), including the requirement that such containers and documents be clearly marked "For Export Only." For further information see section 205.307(c).

Accreditation of Certifying Agents

Posted 1/2/03

The National Organic Program (NOP) has been asked a number of questions regarding how to fulfill the requirement of section 205.501(a)(15)(ii) of the National Organic Standards (NOS) that states certifying agents must submit to the Administrator a copy of a list, on January 2 of each year, including the name, address, and telephone number of each operation granted certification during the preceding year. Based on the questions asked by certifying agents, the NOP provides the following responses to clarify section 205.501(a)(15)(ii):

Q: Does the NOP favor a method of transmission for the lists (i.e. fax, e-mail, US mail, FedEx, or a combination of these)?

A: NOP prefers to receive the lists via e-mail at Arthur.Neal@usda.gov

Q: Is there a particular format preferred by the NOP for the lists? (Word document, Excel spreadsheet, data base listing, etc.)?

A: When submitting lists via email, NOP prefers to receive them in Excel Spreadsheet format.

Q: Does the NOP want the mailing address for the operators, or the physical location of the operation?

A: NOP requires both addresses, if applicable.

Q: The NOS requires that the lists should show the operations "granted certification during the preceding year". Should the lists only contain new operations granted certification, or all operations, including certification renewals?

A: Because this is the initial year for full implementation of the NOS, the NOP requires a complete list of all organic operations certified by the certifying agent as of midnight, December 31, 2002. This would include all operations going through an annual update.

Q: The NOS only asks for the name, address, and telephone number of the operations. Does the NOP also want the categories for which the operations are certified?

A: The NOS does not require identification of the categories for which the operations are certified. However, NOP encourages the submission of this information. Such information will be valuable in identifying the size, scope, and characteristics of the organic production and handling industry.

Q: The NOS requires that the lists be submitted "on" January 2. Can the lists be submitted "on or before" January 2?

A: Yes, the lists may be submitted "on or before" January 2; provided they are complete through the end of the calendar year.

Q: How will a National Organic Program audit of a certifying agent be conducted?

A: The National Organic Program only conducts site evaluations and on-site review audits of certifying agents. Auditors conducting the reviews will follow the procedures specified in the International Organization for Standardization (ISO) 10011 guidelines, "Guidelines for Auditing Quality Systems." Accredited certifying agents will be reviewed based on their ability to comply with the national standards.

December 31, 2002

Q: Please discuss fees associated with audits.

A: Through October 21, 2002, applicants seeking initial accreditation under the National Organic Program (NOP) will only be assessed travel expenses. The NOP will absorb all labor charges for accreditation services through October 21, 2002. After October 21, 2002, applicants for initial accreditation and renewal of accreditation will be assessed fees in accordance to sections 205.640 and 205.641 of subpart G, Fees. Fees for certification are assessed by the certifying agent. Such fees must be in compliance with section 205.642 of the national standards.

Q: Does the National Organic Program (NOP) have plans to produce a national list of accredited certifying agents?

A: Yes. The first announcement of accredited certifying agents will not occur until on or about April 21, 2002. At this time, the NOP will provide a list of accredited certifying agents which will be accessible on the NOP website at http://www.ams.usda.gov/nop/CertifyingAgents/Accredited.html or by request through the NOP office at 1400 Independence Avenue, SW; Room 2510 South Building; Washington, D.C.; 20250.

Q: As a result of the conflict of interest provisions in the national standards, what mechanisms, if any, are there for certifying agents to provide compliance information?

A: Section 205.501(a)(8) requires that certifying agents provide applicants with sufficient information to enable them to comply with the OFPA and regulations. Section 205.501(a)(11)(iv) prohibits certifying agents from giving advice or providing consultancy services to certification applicants or certified operations for the purpose of overcoming barriers to certification. In other words, certifying agents must explain the regulations, but they cannot tell producers or handlers how to correct a noncompliance. Additional discussion of this issue may be found on page 80601 of the national standards.

Q: Based on the conflict of interest provisions in the national standards, how does National Organic Program envision the producer obtaining the correct information to achieve compliance?

A: To address barriers to certification and other organic issues particular to the geographical area, certifying agents may sponsor in-house publications, conferences, workshops, informational meetings, field days, or other educational activities for which participation is voluntary and open to the general public. To overcome barriers to certification, producers and handlers may seek out consultancy services from sources such as educational institutions, State cooperative extension, private consultants, and other producers and handlers.

National List of Allowed and Prohibited Substances

Q: Do nonagricultural substances included on the National List of Allowed and Prohibited Substances have to be produced without the use of volatile synthetic solvents? My certifying agent says yes because of the prohibition on the use of volatile synthetic solvents found in section 205.270(c)(2).

A: No. Section 205.270(c)(2) prohibits the use of a volatile synthetic solvent unless included on the National List as an allowed substance. However, synthetic solvents do not have to be on the National List to be allowed in the production of an allowed nonagricultural substance found on the National List. The use of volatile synthetic solvents in the production of allowed nonagricultural substances included on the National List is considered approved through the materials review process, unless otherwise stated through an annotation to the approved substance. (Example: § 205.605(a)(9), Flavors -- nonsynthetic sources only and must not be produced using synthetic solvents and carrier systems or any artificial preservative.)

Use of Chlorine in Organic Handling Operations - 7 C.F.R 205.601(a)(2), 205.603(a)(3), and 205.605(b)(9) provides for the use of chlorine materials as algicides, disinfectants and sanitizers in organic crop, livestock and handling operations. The annotation on the use of chlorine materials restricts the residual chlorine levels in the water to the maximum residual disinfectant limit under the Safe Drinking Water Act. This limit is currently established by the Environmental Protection Agency (EPA) at 4 mg/L for chlorine. The National Organic Program has received a number of questions regarding the use of chlorine in certified operations and the sampling protocol to be used by accredited certifying agents (ACA) in monitoring the maximum residual disinfectant limit for chlorine materials. The following Q and A's are designed to clarify these issues.

Q. As an ACA, at what point in crop, livestock or handling operations should I monitor for the maximum residual disinfectant limit?

A. ACA's must monitor the discharge or effluent point to ensure that certified operators are meeting the 4 mg/L limit as set forth by the Safe Drinking Water Act.

Q. As a crop, livestock or handling operation, am I restricted to use chlorine at the maximum residual disinfectant limit specified under the Safe Drinking Water Act, currently 4 mg/L, at the beginning of the wash/rinse water cycle?

A. No. Levels of chlorine used to prepare water to be used to disinfect/sanitize tools, equipment, product or food contact surfaces may be higher than 4 mg/L and should be at levels sufficient to control microbial contaminants. Therefore, chlorine use at the beginning of the applicable water cycle in an organic operation is not limited to 4 mg/L.

Q: Can nonagricultural substances not appearing on the National List of Allowed and Prohibited Substances be used as ingredients in or on a product labeled as "made with organic (specified ingredients or food group(s)).?"

A: A "made with organic (specified ingredients or food group(s))" product must, in accordance with section 205.105(c) of the Final Rule, be produced and handled without the use of nonagricultural substances used in or on processed products, except when the nonagricultural substances are included in section 205.605 of the National List of Allowed and Prohibited Substances. Accordingly, the reference to nonorganic ingredients in section 205.301(c) refers to agricultural ingredients only and should not be construed to include nonagricultural ingredients.

To further clarify the Department's intent, a "made with organic (specified ingredients or food group(s))" product must contain at least 70 percent organic agricultural ingredients that have been produced without the use of:

1. Synthetic substances unless the substances and their use are allowed under section 205.601 or section 205.603 of the National List of Allowed and Prohibited Substances.
2. Nonsynthetic substances prohibited under section 205.602 or section 205.604 of the National List of Allowed and Prohibited Substances.
3. Nonagricultural substances unless the substances are allowed under section 205.605 of the National List of Allowed and Prohibited Substances.

Additionally, the remainder of the ingredients in a "made with organic (specified ingredients or food group(s))" product (up to 30 percent) may include:

1. Nonagricultural products listed in section 205.605 of the National List.
2. Nonorganically produced agricultural products, raw or processed, that have been produced using synthetic, nonsynthetic, and nonagricultural substances without regard to sections 205.601 through 205.605 of the National List of Allowed and Prohibited Substances, except that the use of excluded methods, sewage sludge, and ionizing radiation are prohibited. Nonorganically produced agricultural products listed in section 205.606 of the National List of Allowed and Prohibited Substances must comply with the restrictions placed on that product by section 205.606.

Q: Will there be a generic list of allowed naturals?
A: No, there will not be a generic list of allowed natural materials, because all naturals are allowed unless prohibited on the National List.

Q. What is the “maximum residual disinfectant level?”
A. “Maximum residual disinfectant level” is a term defined by the Environmental Protection Agency (EPA) as the highest level of a disinfectant allowed in drinking water. This level is currently established by EPA at 4 mg/L for chlorine. Practically applied under the National Organic Standards, the term “maximum residual disinfectant level” refers to the chlorine level of the waste water at the discharge or effluent point.

State Organic Programs

Q: Can a State request review of a proposed State Organic Program (SOP) before all the statutory and regulatory rules have been changed? For instance, can a State specify its intent to adopt legislation or rules in its request for approval, rather than having those statutes or rules finalized?

A: A State has two options: (1) It can submit a final SOP to USDA for approval; or (2) It may submit to USDA, a draft SOP for review prior to finalization of statutes, regulations, or procedures. When this option is used, USDA will provide feedback to the State regarding any changes necessary to receive approval. A State can then use the feedback to finalize its SOP. The SOP would, then, be submitted to USDA for approval.

Q: Can a State require registration of all certifying agents operating within the State?

A: Yes; provided that the registration program does not discriminate against the certifying agents.

Q: Can a State with an approved State Organic Program (SOP) review the operations of certifying agents operating within the State?

A: Only States with approved SOPs may review and investigate complaints of noncompliance with the Act or regulations concerning accreditation of certifying agents operating in the State (See section 205.668(c)).

Q: Can a State require certifying agents operating within the State to comply with the approved State Organic Program's (SOP's) more restrictive requirements?

A: Accredited certifying agents operating within a State with an approved SOP must comply with the SOP's more restrictive requirements. A violation of the State's approved more restrictive requirements is a violation of the National Organic Standards (NOS). Only the USDA may take enforcement action against an accredited certifying agent for violating the NOS. States with approved SOPs may review and investigate complaints of noncompliance and report their findings to the NOP for enforcement action (See section 205.668(c)).

Q: When is a State considered to have a State Organic Program (SOP)?

A: A State is considered to have an SOP when it receives USDA approval. To receive approval, a proposed SOP must meet the National Organic Standards, receive approval for any identified more restrictive requirements, and contain noncompliance, mediation, and appeals procedures that meet the requirements of USDA. The basis for the recommendation to approve or disapprove will be based on compliance with the requirements of Sections 205.620 through 205.622, 205.661 through 205.663, 205.668, and 205.680 and 205.681 of the NOS (7 CFR 205.620-.622, 205.661-.663, 205.668, and 205.680-.681).

Q: Who is responsible for handling a U.S. District Court appeal of a State's final decision?

A: The appeal is on a State action, and the State is responsible for defending its action.

Q: What discretion, if any, does a State Organic Program (SOP) have in handling an appeal of an enforcement action initiated by an accredited certifying agent?

A: None. Approved SOPs are required to handle all appeals of enforcement actions initiated by certifying agents against certified operations operating within the State.

Q: Can a State Organic Program (SOP) include additional enforcement provisions, such as authority to issue cease and desist orders, obtain injunctions to stop the sale of noncompliant products, or summarily suspend certifications?

A: A State may request approval of an SOP that includes more restrictive enforcement requirements that do not deny due process.

Q: Will an approved State Organic Program (SOP) be responsible for revoking the certification of a certified operation based in foreign countries?

A: When a certified entity operates in multiple locations and one of those locations is within the jurisdiction of an SOP, the State is responsible for enforcing compliance by that portion of the certified operation.

Q: If a certified producer or handler fails to meet the continuation of certification requirements and contests a proposed revocation of certification, can a State with a State Organic Program (SOP) stop that entity from selling its product as "organic" prior to completion of the appeals process?

A: A stop sale before completion of the appeals process would be a denial of due process.

Q: Is it possible for organic growers to be in compliance with the National Organic Program if their State does not create a State Organic Program.

A: Yes. A State does not have to create a State Organic Program or become an accredited certifying agent for organic growers within the State to be in compliance with the requirements of the National Organic Program.

Q: Please explain the options available to a State considering the possibility of creating a State Organic Program (SOP) under the National Organic Program (NOP).

A: *Option 1*: A State may choose not to establish an SOP or provide certification services under the NOP. Under this option, organic growers may seek organic certification by any certifying agent accredited under the NOP. The State would not be responsibly connected to enforcement of the NOP. Enforcement would be shared jointly by the NOP and the certifying agent.

Option 2: A State may choose to only provide certification services under the NOP. Under this option, the State would have to apply for accreditation. As an accredited certifying agent, the State would be responsible for conducting certifications, enforcing the production and handling standards of the NOP, and maintaining compliance with other applicable NOP regulations. USDA will be responsible for oversight of the State certifying agent. Organic producers and handlers within that State could choose to be certified by the State or any other accredited certifying agent.

Option 3: A State may choose to establish an SOP. Under this option, all organic producers or handlers in the State would have to be certified according to the SOP, which would include the requirements of the NOP and the more restrictive provisions unique to that State and approved by the USDA. The State would assume enforcement responsibility, within its borders, for the requirements in the national standards and its SOP (§205.620(d)). However, the State may not initiate proceedings to suspend or revoke the accreditation of any certifying agent accredited by USDA. Suspension or revocation of an accredited certifying agent may only be pursued by the USDA. Organic producers and handlers may seek and obtain organic certification from any certifying agent accredited under the NOP.

Option 4: A State may choose a combination of Options 2 and 3.

Exporting Organic Products

Q: When does organic product have to be labeled "For Export Only"?

A: A product must be labeled "For Export Only" when that product has been produced in the United States to an organic standard other than the National Organic Standards (NOS).

Q: Can domestically produced organic product that meets the NOS and an international organic standard make a claim regarding the international standard on the PDP without violating the NOS?

A: Yes. Truthful label claims are allowed under the NOS.

References

[1] 7CFR205.2.

[2] USDA, Economic Research Service. *Recent Growth Patterns in the U.S. Organic Foods Market.* Report No. AIB777. September 2002. Available online at [http://www.ers.usda.gov/publications/aib777/].

[3] For a detailed graphic representation of the distribution of organic crop and pasture acreage and number of operations per state, see [http://www.ers.usda.gov/Amberwaves/Feb03/ pdf/indicators.pdf].

[4] USDA. Economic Research Service. Price premiums hold on as U.S. organic produce market expands. Report VGS-308-01. May 2005. Available online at [http://www.ers.usda.gov].

[5] Ibid.

[6] Ibid.

[7] Ibid.

[8] Ibid.

[9] Farms and handling operations that sell less than $5,000 a year in organic agricultural products are exempt from certification; however, these producers and handlers must abide by the national standards for organic products and may label their products as organic.

[10] A July 2005 audit by the USDA Office of Inspector General (OIG) states that USDA had 41 accredited certification agents in foreign countries. The report also found that as of August 2004, AMS had negotiated only one equivalency agreement with a foreign country (Japan). The OIG recommends implementation of internal operating procedures “to assure that the NOP is achieving its intended objectives to ensure that organic

products meet consistent, uniform standards." The audit report is available online by going to [http://www.usda.gov/oig/audits.htm] and following the links from "View Audit Reports and News Releases," to the listing by "Agency Subject of Report." It is under the Agricultural Marketing Service, July 2005.

[11] The CCC is a wholly owned government financing institution for USDA agencies that administer mandatory programs, such as the farm commodity price and income support programs for wheat, cotton, rice, and certain other crops; agricultural export subsidies; and certain conservation and trade programs. CCC funds are considered mandatory funds that must be made available for the purposes authorized. In practice, however, appropriators sometimes prohibit or place restrictions on funding for mandatory programs in the annual appropriations bill. A July 2005 audit of the NOP by USDA's Office of Inspector General (OIG) found that AMS has not been properly documenting its procedures for maintaining and controlling cost-share programs, which makes it difficult if not impossible to monitor the use of funds. AMS officials have promised to implement proper procedures by the end of FY2005.

[12] Information on Integrated Organic Program research grants are available on the Cooperative State Research, Education, and Extension Service website at [http://www.csrees.usda.gov/fo/funding.cfm].

[13] ERS published a paper in August 2005 entitled, "Market-Led Growth vs. Government-Facilitated Growth: Development of the U.S. and EU Organic Agriculture Sectors." Available online at [http://www.ers.usda.gov/Briefing/Organic/]. Congress has not appropriated funds to date for a National Agricultural Library International Organic Research Collaboration.

[14] USDA, Agricultural Marketing Service. *Organic Feed for Poultry & Livestock: Availability and Prices.* Available online at [http://www.ams.usda.gov/nop/prodhandlers/feedstudyjune2003.pdf].

[15] Available online at [http://www.ams.usda.gov/nop/TodaysNews.html]; see August 24, 2005.

[16] USDA, Office of Inspector General. *Audit Report: Agricultural Marketing Service's National Organic Program.* The audit report is available online by going to [http://www.usda.gov/oig/audits.htm] and following the links from "View Audit Reports and News Releases," to the listing by "Agency Subject of Report." It is under the Agricultural Marketing Service, July 2005.

[17] Notice was published in the *Federal Register* on August 13, 2003 (68 FR 48334).

[18] Available online at [http://www.ams.usda.gov/nop/CertifyingAgents/ANSIReport.pdf].
[19] "Organic Milk Debate; Dairies dispute 'organic' values; Ex-hippie farmers contest practices of big producers," *Chicago Tribune*, January 10, 2005. See [http://www.organicconsumers.org/organic/milk011105.cfm].
[20] Minutes of February 28-March 3, 2005, NOSB meeting available online at [http://www.ams.usda.gov/nosb/meetings/meetings.html].
[21] Minutes of August 15-17, 2005, NOSB meeting available online at [http://www.ams.usda.gov/nosb/meetings/meetings.html].
[22] "NOSB Livestock Committee Recommendations for Guidance; Pasture Requirements for the National Organic Program," National Organic Standards Board, March 2, 2005. Available online at [http://www.ams.usda.gov/nosb/meetings/meetings.html].
[23] *Harvey v. Veneman*, 396 F.3d 28 (1st Cir. 2005).

In: Organic Agriculture in the U.S.
Editor: Alison J. Wellson, pp. 55-69
ISBN: 978-1-60021-305-2

Chapter 2

ORGANIC PRODUCE, PRICE PREMIUMS, AND ECO-LABELING IN U.S. FARMERS' MARKETS

Amy Kremen[1], Catherine Greene[1] and Jim Hanson[2]
[1]College of Agriculture and Natural Resources,
The University of Maryland, USA
[2]Resource Economics Division, Economic Research Service

ABSTRACT

The number and popularity of farmers' markets in the United States have grown concurrently with organic production and consumer interest in locally and organically produced foods. This research describes the significance of farmers' markets as market outlets for many organic farmers, and recent shifts in relationships between organic growers, market managers, and customers. Market managers in more than 20 States answered questions by phone pertaining to the 2002 market season. Their responses provide insight into recent grower, manager, and customer decision making and attitudes about foods advertised and sold as organic at farmers' markets. Specific areas of interest include participation of organic farmers in markets, demand for organic products, advertising strategies used to highlight organic products, and price premiums charged for organic products.

INTRODUCTION

Farmers' markets have been growing in number and popularity over the past decade, providing valuable opportunities for thousands of full- and part-time farmers. While the volume of produce sales via farmers' markets is small—less than 2 percent of U.S. sales overall (Handy et al.)—the number of farmers' markets in the U.S. has increased by 79 percent since 1994 to more than 3,100 in 2002. Consumers representing a wide range of socioeconomic backgrounds are demonstrating growing interest in and support for local agriculture (Stephenson and Lev).

Participation by growers at farmers' markets depends on several factors, including a market's popularity, location, and management. Customer participation depends primarily on a market's location, since most customers tend to shop at markets close to where they live (Brown). Ultimately, markets that attract and retain vendors and shoppers are those that satisfy the needs of both (Brown). Market success can depend on how attuned managers are to customer demand and farmers' interests, and on how well they can satisfy these needs.

Previous studies indicate that freshness, high quality, fair pricing, pleasant social interaction with farmers and market shoppers, and locally grown foods are attributes commonly sought by customers attending farmers' markets (Lockeretz, Brown). Vendors at farmers' markets also frequently provide heirloom and specialty produce varieties, hand-crafted products, and other items valued by customers but difficult to mass-produce and are rare at other markets (Lyson et al.). This study shows that many customers at farmers' markets also appreciate having direct access to farmers that use ecologically sensitive agricultural techniques—such as organic production methods—on their farms.

Research on the marketing habits of organic growers indicates that they tend to use direct marketing outlets, such as farmers' markets, much more often than the general U.S farmer population (Walz, Fernandez et al.). Although the marketing channels for organic food have expanded in recent years, farmers' markets remain an accessible, low-cost, and flexible sales outlet for many organic producers. Insufficient infrastructure—a lack of established purchasing, storage, and distribution channels, for example—may still hinder growers interested in expanding or coordinating their production to serve other local market outlets (Slama, Birkerts). Several market managers interviewed for this study indicated that organic farmers routinely play a significant role in creating farmers' markets, and often serve as manager or board member.

For this study, farmers' market managers in more than 20 States answered questions by phone pertaining to the 2002 market season. Market managers were interviewed in late 2002 and early 2003 about participation by organic growers and customer demand for organic products. Questions about the 2002 market season (14 total) were modeled on the market manager questionnaire being used in an ongoing study of U.S. farmers' markets by USDA's Agricultural Marketing Service. The AMS study, initially conducted in 1994 and repeated in 2000, surveys all farmers' market managers in the United States, and has shown that farmers' markets are "a growing marketing tool for farmers" and that "customers can benefit from direct contact with the producers of their food" (Payne).

The AMS study collects a wide range of information on farmer and customer demographics and market characteristics. This study focuses more narrowly on the role of farmers' markets as a marketing tool for organic growers and on current customer attitudes about organic products at these markets.

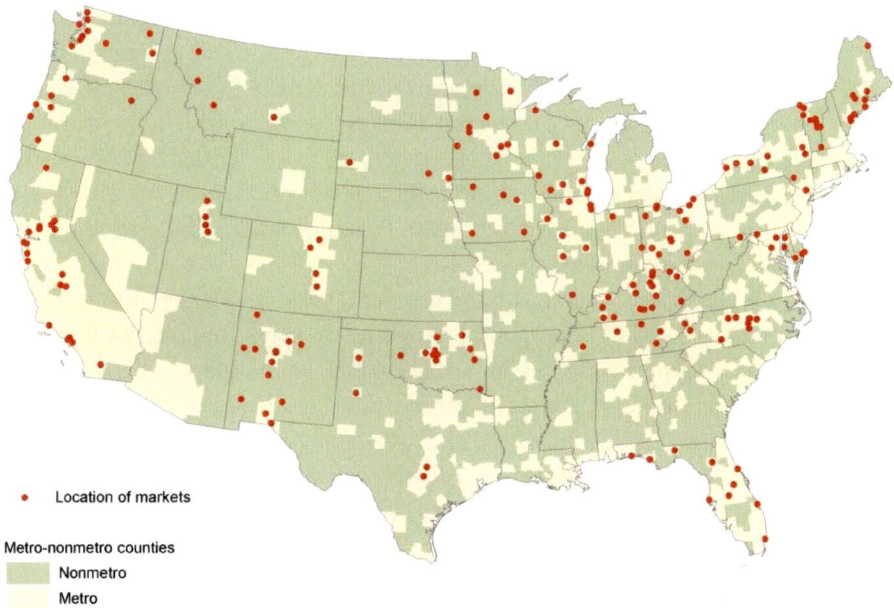

Map 1. Distribution of 210 U.S. farmers' markets interviewed, 2002

The group of market managers interviewed was selected with guidance from AMS-appointed State farmers' market representatives who maintain lists of managers and markets. State representatives were asked to provide a list of approximately 15 markets that would represent the diversity in size, "feel," and location of their State's markets. Interviews were conducted with 210 market managers located in 23 States and Washington, DC (map 1). Although market sampling was not random, the data generated provide useful information for understanding the current role of organic growers at markets within a wide range of communities. The data collected in this study for many market characteristics—including market type, average levels of farmer attendance, and customer distances traveled to market—correspond closely to data collected for those characteristics by AMS in its much larger study of U.S. farmers' markets.

Organic Farmers Participate Widely in U.S. Farmers' Markets

Organic growers participated at more than four-fifths of the 210 markets in this study, and represented approximately one-third of regularly attending farmers during the 2002 season (table 1). Such broad participation highlights the disproportionately high use of farmers' markets as a sales outlet by organic growers. While only one or two organic growers participated in about half of the markets with organic growers, 11 percent of these markets had more than ten participating organic growers (table 1). Many of these farmers had already acquired third-party organic certification, or had developed alternative eco-labels to promote their products, even though Federal rules requiring organic certification had not yet taken effect (see box).

The data collected in this study are not conclusive about whether organic growers are more likely to attend urban (metro) or rural (nonmetro) farmers' markets. Organic farmer participation was sometimes limited or negligible at several large, highly popular urban markets, but was significant at several smaller metro and rural markets, a few of which had exclusively organic vendors.

Managers were able to provide general estimates of gross sales (not necessarily limited to sales at the market) of the organic farmers participating at the markets in this study. Nearly 90 percent of the certified organic growers, and slightly over half of the noncertified growers, were likely to be grossing more than $5,000 per season from their farming operations. According to managers, market sales of organic farmers were comparable to or better than sales of other growers in the markets in this study.

Managers at several smaller markets in rural areas commented that some growers—full-time farmers as well as hobby gardeners—are developing an interest in organic production, even though they may be marketing in areas where customer demand is low or where experience with organic production among farmers and extension is limited. According to managers, this budding farmer interest is driven by personal, financial, and environmental motivations, as well as a desire to position themselves for a burgeoning customer base.

Table 1. Organic grower participation in 210 U.S. farmers' markets,2002

Organic farmer participation	
Number of farmers in 210 farmers'markets	4,681
Number of participating organic farmers	1,416
Markets with organic farmers (percent)	82
Number of organic farmers present (percent):	
1-2 growers	46
3-10 growers	43
11-40 growers	11

Most managers indicated that organic farmer participation has either increased or remained constant since the 2000 season, usually for two reasons. First, more farmers than in previous years were choosing to sell directly at farmers'markets while opting simultaneously to farm using organic methods. Second, the rapid growth in size and popularity of many markets was a drawing force for organic farmers, especially in situations where customer demand was compelling managers to search for additional organic growers.

Participation at the markets by certified organic growers had generally increased by one or two farmers within the previous 3 years, while participation by non-certified organic growers rose by one or more farmers yearly. Ten percent of markets studied either gained organic growers for the first time within the previous 3 years or had increased participation by organic growers. Markets with several organic growers often had a strong tendency to attract additional organic growers. Most markets without any participation by organic growers did not tend to gain any organic growers within the previous 3 years.

Managers indicated that customer interest in organic products at those markets was generally low.

Just over half of the managers indicated that the participation level of organic growers at their markets remained constant over the previous 3 years. In most cases, organic growers were reportedly successful at their markets and were likely to continue participating. At some markets where limited stall space was preventing the addition of other vendors, managers were planning to make room for organic growers to help meet strong customer demand. Some markets simply lacked organic farmers nearby who were not already committed to other markets.

In markets where the participation of organic growers was decreasing, managers explained that their organic farmers were moving on to bigger or wholesale markets, or were ceasing to farm for personal reasons not related to age. Managers often noted that organic farmers seemed younger than conventional farmers, and that they tended to drop out of markets for age-related reasons less frequently than conventional farmers.

Organic Standards and Certification

Congress passed the Organic Foods Production Act of 1990 to establish national standards for organically produced commodities, and USDA promulgated final rules that implemented this legislation on October 21, 2002. These regulations require that all organic growers and handlers (including food processors) be certified by a State or private agency accredited under the uniform standards developed by USDA, unless the farmers and handlers sell less than $5,000 a year in organic agricultural products. Retail food establishments that sell organically produced agricultural products but do not process them are also exempt from certification.

The national organic standards address the methods, practices, and substances used in producing and handling crops, livestock, and processed agricultural products. Although specific practices and materials used by organic operations may vary, the standards require every aspect of organic production and handling to comply with the provisions of the Organic Foods Production Act.

The labeling requirements under the national standards apply to raw, fresh products and processed foods that contain organic ingredients and are based on the percentage of organic ingredients in a product. Agricultural products labeled "100 percent organic" must contain (excluding water and salt) only organically produced ingredients. Products labeled "organic" must consist of at least 95 percent organically produced ingredients. Products labeled "made with organic ingredients" must contain at least 70 percent organic ingredients. Products with less than 70 percent organic ingredients cannot use the term organic anywhere on the principal display panel but may identify the specific ingredients that are organically produced on the ingredients statement on the information panel. In a processed product labeled as "organic," all agricultural ingredients must be organically produced unless the ingredient(s) is not commercially available in organic form. The USDA organic seal—the words "USDA organic" inside a circle—may be used on agricultural products that are "100 percent organic" or "organic." A civil penalty of up to $10,000 can be levied on any person who knowingly sells or labels as organic a product that is not produced and handled in accordance with these regulations.

DEMAND FOR ORGANIC PRODUCTS

Market managers reported that demand for organic products was strong in nearly 40 percent of the markets that had organic farmers, medium in 47 percent of these markets, and low in only 13 percent. Farmers' markets near major urban areas tended to have strong demand for organic products (map 2). Market communities near universities and other centers for higher education, near certain religious communities, or near facilities for holistic health care were also likely to exhibit strong demand. Markets with concentrations of organic growers, or one or two large-scale organic growers, tended to be associated with strong demand for organic products, regardless of setting. In rural areas, some markets had strong customer demand for organic products, according to managers, because they provided access locally to well-priced, fresh, organic foods otherwise only available in cities or towns some distance away. According to managers, customers with strong demand for organically grown products were most likely to exhibit interest in the social and environmental issues within agriculture, such as the relationship of agriculture to human and animal health, sustainable development, water scarcity, environmental pollution, and wildlife protection.

Growing interest in organic products led several managers to search for additional organic farmers to participate at their markets. Managers in California and Florida, in particular, indicated that there were not enough organic growers available to supply the quantity and variety of organic products needed to meet customer demand. Managers in many other States—including Colorado, Illinois, Maryland, Minnesota, North Carolina, New York, Tennessee, Utah, and Wisconsin—indicated that favorable opportunities for organic growers existed at their markets.

According to managers, customer interest in organic products has developed through customer contact with organic growers at farmers'markets, other direct marketing relationships, and influences such as increased volume and visibility of organic products in other market venues. Managers indicated that it was typical for organic farmers to exert a strong influence at their markets by bringing a wider range of specialty products than other market vendors. Organic farmers have historically played the role of the reliable, "cornerstone" vendors at some markets, which was also influential in generating demand for organic products, according to managers.

Managers noted that customers did not usually decide to buy from organic farmers simply because their products were organically grown. According to managers, organic farmers tended to excel at markets if they also satisfied customer desire for other attributes, such as consistently high product quality, access to specialty crop varieties, and excellent customer service. For example, one manager from Liberty, New York, indicated, "Partly because of the wide variety and sheer beauty of their stands, organic growers do well at my market." Managers in Illinois explained that "[conventional] farmers will typically bring peas, beans, cabbage, and squash, while organic growers will bring lettuce mixes, sprouts, spinach, cauliflower, and edamame (edible soybeans)" and "different things like parsnips and leeks."

Managers indicated that organic farmers are likely to converse with customers who ask how to prepare specialty food items, or inquire about organic farming techniques and associated costs of production. Such communication, they noted, can foster relationships between farmers and customers that may persist, especially if a customer is also satisfied with a farmer's product quality. One manager at an all-organic market in Vermont explained that "the most important thing is establishing a relationship between the customers and the growers. New growers take 3 to 4 weeks to gain solid interest from customers who are waiting to see if a grower is going to be reliable with product quality and attendance…That connection is more important than [a] label."

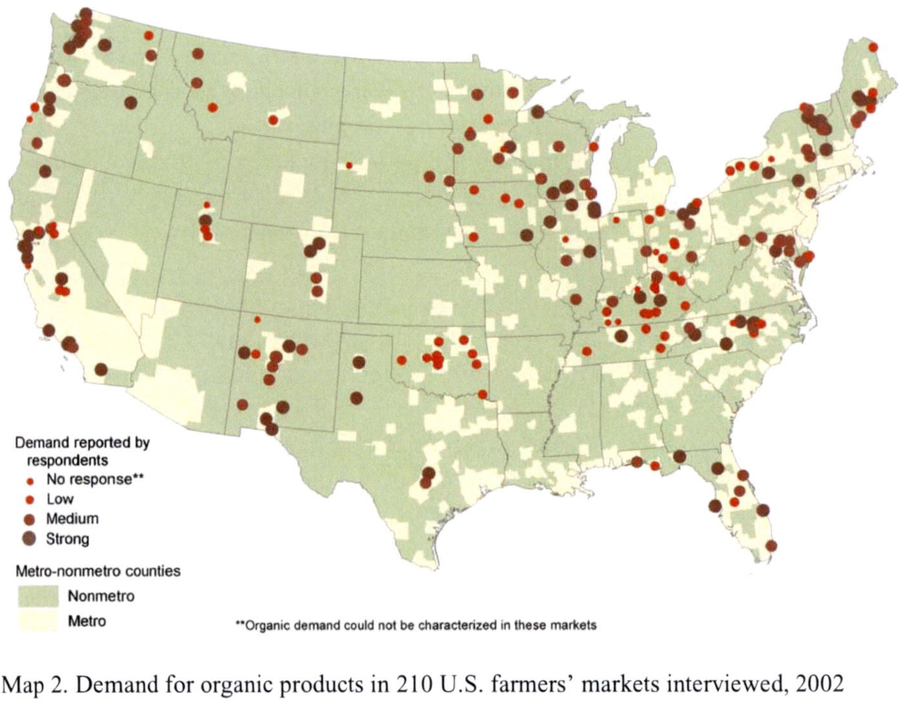

Map 2. Demand for organic products in 210 U.S. farmers' markets interviewed, 2002

At some markets in this study, customer demand for organic products has had a strong influence on some farmers who recently decided to transition to organic farming or lower-input farming practices. For example, at a few markets in western Washington, several large-scale vegetable and fruit farmers who were switching from wholesaling to direct marketing at farmers'markets found that transitioning to organic production could enhance their marketing potential. "'Are you organic?'and 'Are you certified?'are the most frequently asked questions" at one market in Seattle, according to the manager. The influence of customer demand for organic products on growers' production methods was most prominent at markets in Washington and New Mexico.

Markets characterized in this study as having "medium" levels of demand for organic products had a small, often fast-growing, group of regular customers for organic products. Half of these markets were in towns or cities of less than 20,000 people, and had about half as many participating farmers as the average market in this study. A majority of the remaining markets with medium levels of demand were in cities of less than 100,000 people. Tourists visiting some markets in resort/vacation areas in Minnesota, New Mexico, and Washington contributed significantly to customer demand for organic products.

Markets where customers and farmers had little interest in—or negative opinions about—organically grown products were often in areas that managers described as being rural or still in early stages of organic food awareness. These markets had higher attendance by customers with limited incomes, compared with other markets. While customers had not developed much interest in organic products at these markets, several managers said that their customers had shown some interest in foods produced using fewer inputs. A few managers indicated that customers were reluctant to purchase organic products either due to previous negative experiences or expectations with regard to the quality or prices of organically grown products, or because organic growers had attended their markets too sporadically. Some market managers, particularly in the South, indicated that their farmers perceived organic farming as difficult, labor-intensive, and uneconomical due to pressures of weeds, pests, and diseases.

Education—about farming systems, sustainable agriculture, and how to prepare fresh food—was perceived by managers as the element most likely to generate additional interest in organic products and production from customers and conventional farmers. Several managers also suggested that relieving the short supply of organic products at their markets would likely stimulate interest and demand. For example, as one manager from New York explained, "People tend to think that everything at the market is organic, and when they find out that everything isn't, they are disappointed and want more. It's easier to make the market grow with more supply [of organic products]."

ECO-LABELS AND DIALOGUE EDUCATE MARKET CONSUMERS

Farmers'markets often receive coverage by the popular press, which helps offset small advertising budgets. Most managers interviewed for this study said they used very simple, low-cost techniques—such as articles, flyers, public service announcements for radio broadcast, websites, and notices for church and nonprofit newsletters—to attract farmers and customers to their markets. Managers generally promoted organic farmers and products only when organic farmers comprised over half of the vendors. Managers were reluctant to risk alienating their conventional growers with organic advertising or to inadvertently create a false impression that all products at their markets were organic. A small group of managers in nonmetro areas noted that advertising organic products and farmers would not be particularly effective, either due to negative perceptions or because the term organic was not "part of the community's vocabulary."

Markets with a strong history of organic farmer participation were more likely to advertise organic products and growers, as were those markets supported by nonprofit organizations. At some markets, where organic growers were a recent addition or where customer demand for organic products had grown significantly, managers were considering the use of some organic advertising for the 2003 season. Managers used an array of techniques—including color-coded market banners and price tags, detailed brochures, and website profiles of growers—to help orient and inform customers about certified organic, transitional, "pesticide-free," and other farming methods. Some techniques used by managers to promote their market, such as popular "Chef-at-Market" programs, often highlighted organic products and farmers.

Though some managers guide their farmers on how to successfully advertise their products, individual farm-ers'efforts vary. Some managers expressed surprise—or frustration—that organic farmers did not emphasize their organic attribute more, while others explained that only recently had shifts in customer awareness and interest in organic products made it more advantageous for farmers to advertise their organic growing practices. For example, one manager from Illinois indicated that her "customers feel like they can quit being afraid of organic, with more and more media being positive about it." Managers at a few markets, particularly in rural areas, indicated that organic product advertising was unlikely or still a few years away.

According to managers, talking was the main means of advertising used by non-certified organic growers to inform customers about products in 2002, while certified organic growers were more likely to use some kind of signage to promote their products. A range of signage types was used by growers, including individual labeling of products with the word "organic," displaying banners or plaques from certification agencies and farmer associations, and incorporating the word "organic" in farm names. Managers observed that some conventional producers have been using signs to indicate which crops, like lettuce and some fruit crops, had been grown without synthetic inputs.

Slightly under half of the organic growers in the markets studied during the 2002 market season were certified. Federal organic regulations requiring certification of products labeled organic were implemented after the 2002 market season had ended (see box, p. 5). Even prior to implementation of these regulations, about 15 percent of the markets studied did not allow noncertified organic produce to be labeled as organic. Other labels being used by organic farmers and others using alternative production methods were:

- "Chemical-free," "no chemical fertilizers," and "no harmful insecticides"
- "Natural growing conditions" and "naturally-grown"
- "Healthy farming practices used"
- "No-spray" • "Sustainably grown"
- "Authentic"
- "Free-range," "pastured," "grass fed," and "no antibiotics or hormones" • "Organically inclined"
- "Transitional" • "Wisconsin organic grown" and other local foods labels
- "Good bugs at work here"
- "Ask me how I grow this."

Organic farmers'advertising efforts also included picture displays of farms, recipes and business cards, websites, informational brochures, and farm tours.

Table 2. Advertising techniques used by organic growers in 210 U.S. farmers markets, 2002

Organic advertising techniques[1]	No. farmers
Talking (interaction)	187
Signs and labeling advertising organic methods	129
Posting of organic certification banners/plaques	100
Brochures, newsletters, business cards, and recipes	61
Photos	21

[1] Some farmers use multiple techniques.

Organic Price Premiums Differ Across Markets

According to managers, prices for organic products were determined both casually and formally by individual farmers or by organic farmers as a group, sometimes with the help of pricing services such as AMS *Market News*, State price reports, and agricultural publications such as *Growing for Market*, which publish some information on wholesale prices for organic crops.

The use of price premiums for organic crops depended on seasonality and individual farmer choice. At about 40 percent of the markets where they participated, organic farmers did not request premium prices. Organic farmers charged a premium for their products, often modest, at about half of the markets where they participated. In some markets, there was no basis for comparison because the markets were all-organic or the organic and conventional growers sold different items. According to managers, organic farmers tended to set significant premiums only for products they perceived as being of exceptionally high quality, especially rare, or for markets in upscale communities. Premiums charged for poor quality product were unusual, and poorly received by farmers'market shoppers.

Organic farmers were more likely to ask for premiums when attending large markets where organic farmers represented half or more of the growers. Organic farmers participating at these markets generally had enough production to supply the market with a size-able volume of product, and were likely to be full- or part-time growers earning a significant portion of their household income from their farm operations.

Certified and noncertified organic growers were equally likely to charge premiums for organic products at these markets.

Based on this information about the rarity and modesty of price premiums, some customers'reluctance to purchase organic foods may have less to do with actual prices than with expected prices for organic products. According to one manager in Maine, for example, "people have an image that organic is more expensive and may pass you by if you advertise too much."

Conclusion

Farmers'markets are providing an important marketing outlet for many farmers, including organic farmers, who want to generate substantial income from high-value crops grown on minimal acreage. Fair pricing, product quality and consumer trust are essential to the overall success of both the conventional and organic farmers participating in these markets. It is not unusual for consumers to visit farmers'markets thinking that most or all of the fresh produce sold there is organically produced. Farmers'markets provide a natural arena where organic farmers can educate consumers about what distinguishes their crops from the others being offered.

Information collected from farmers'market managers for this study suggest that demand for organic products is substantial and growing in many U.S. farmers'mar-kets. Demand for organic products and participation by organic growers at markets may be dampened by:

- Limited numbers of local organic farmers available to sell at markets,
- Limited awareness and interest of some consumers and farmers in organic production systems, and
- Negative perceptions of organic products or organic product pricing.

Many producers, even large-scale fruit and vegetable farmers, have become interested in incorporating organic production systems or direct marketing channels into their farming operations. Further investigation of direct marketing could shed light on the significance of net income derived from organic product sales at these markets, identify ways that organic growers acquire and use their direct marketing expertise, and clarify other poorly understood aspects of this marketing channel.

Acknowledgments

The authors gratefully acknowledge the comments and reviews of Kitty Smith, Joy Harwood, and Keith Wiebe (ERS); John Love (WAOB); Barbara Robinson (AMS); and Gail Feenstra (University of California, Davis) and Garry Stephenson (Oregon State University, Corvallis). The authors also thank Chris Dicken for excellent geographic information system support and Dale Simms for valuable editorial assistance. We are also indebted to the farmers' market managers and State representatives who shared their observations and expertise with us for this project.

References

Birkerts, Erik G. 2000. "The Local Market Opportunity for Organic Produce:An Objective Business Analysis," Prairie Partners Group, LLC.

Brown,Allison. 2002. "Farmers market research 1940-2000:An inventory and review." *American Journal of Alternative Agriculture*,Vol. 17, No. 4,167-176.

Duram, Leslie A. 1999. "Factors in organic farmers' decision making: Diversity, challenge, and obstacles," *American Journal of Alternative Agriculture*, Vol. 14, No. 1, 2-10.

Dimitri, Carolyn and Nessa Richman. 2000. *Organic Food Markets in Transition*, Henry A Wallace Center for Agricultural and Environmental Policy, Winrock International.

Federal Register. 2000. *National Organic Program; Final Rule*. Table 1- U.S. Organic Product Sales, 1990-1999,Vol. 65, No. 246, Dec. 21.

Fernandez-Cornejo, Jorge, Catherine Greene, Renata Penn and Doris Newton. 1998. "Organic vegetable production in the U.S.: Certified growers and their practices," *American Journal of Alternative Agriculture*,Vol. 13, No. 2.

Handy, Charles R., Phil R. Kaufman, Kristen Park, and Geoffrey M. Green, 2000. "Evolving Marketing Channels Reveal Dynamic U.S. Produce Industry," *FoodReview*,Vol. 23, Issue 2, U.S. Department of Agriculture, Economic Research Service, May-August.

Lockeretz,William (ed.). 1987. *Sustaining Agriculture Near Cities*. Soil and Water Conservation Society,Ankeny, Iowa.

Lyson, T.A., G.W. Gillespie, Jr., and D. Hilchey. 1995. "Farmers' markets and the local community: Bridging the formal and informal economy," *American Journal of Alternative Agriculture*,Vol. 10, No. 3, 108-113.

Payne, Tim. 2002. *U.S. Farmers Markets-2000, A Study of Emerging Trends*, U.S. Department of Agriculture,Agricultural Marketing Service, Transportation and Marketing Programs, Marketing Services Branch.

Slama, Jim. 2002. *The Land of Organic Opportunity-Steps to Building a Regional Organic Food System Serving Chicago*, Sustain Alliance, Chicago, IL, www.sustainusa.org/local/organic/.

Stephenson, Garry, and Larry Lev. 1998. *Common Support for Local Agriculture in Two Contrasting Cities*, Oregon State University, OSU Extension Service, Oregon Small Farms Technical Report Number 1.

Walz, Erica. 1999. "Third National Organic Farmers' Survey," Organic Farming Research Foundation, Santa Cruz, California.

In: Organic Agriculture in the U.S.
Editor: Alison J. Wellson, pp. 71-95

ISBN: 978-1-60021-305-2

Chapter 3

PRICE PREMIUMS HOLD ON AS U.S. ORGANIC PRODUCE MARKET EXPANDS*

Lydia Oberholtzer, Carolyn Dimitri and Catherine Greene

ABSTRACT

Fresh fruits and vegetables have long been an important component of the organic food sector. Price premiums for organic products have contributed to growth in certified organic farmland and, ultimately, market expansion. This report examines trends in organic prices and market margins for broccoli, carrots, and mesclun mix. The data show that, while organic wholesale price premiums for mesclun mix are narrowing, wholesale and farmgate premiums for broccoli and carrots remain strong.

Keywords: Organic, fresh produce, fruits and vegetables, organic price premiums, organic marketing margins.

INTRODUCTION

Organic produce, once sold only in specialty shops, is currently available in a wide variety of food retail outlets. This new accessibility is the result of many demand- and supply-side forces. Consumer demand for variety, convenience, and

---

* Note : A web data product with monthly organic farmgate and wholesale prices for broccoli, carrots and mesclun mix is available at www.ers.usda.gov/data/ OrganicPrices.

quality in fresh produce—both organic and conventional—has grown rapidly. As a result, suppliers are increasingly introducing new varieties, and retailers now offer many organic fresh produce items year-round. Rising consumer demand for convenient products has boosted sales of organic (and conventional) pre-cut produce, with more packaged and branded products available. In response to the growing popularity of organic products, conventional supermarkets and mass market merchandisers have added organic fruits and vegetables to their shelves.

Organic fruits and vegetables are important in the organic sector because they are "gateway" products, or the first organic products purchased by consumers. Organic gateway products, which also include dairy, nondairy (soy), and baby food products, often steer consumers toward other organic products, such as cereals, snacks, and meat and poultry, and are perceived as important frontline commodities for the industry.

USDA's National Organic Program regulates organic products (see box, "National Organic Standards"). USDA introduced national organic standards on October 21, 2002, 12 years after they were mandated by the Organic Foods Production Act of 1990. U.S. organic standards were set in place to build consumer confidence in organic products, and, in turn, sustain and stimulate growth of the industry as a whole. In addition, many industry members hope that the standards will open markets for U.S. exports of organic products through regulatory equivalency.

It is still too early to gauge some of the effects of the regulations; however, since implementation of the standards, U.S. organic sales have continued to grow at rates maintained over the last decade—20 percent or more per year. The fresh produce sector constitutes the largest sector of the organic industry and thus accounts for a significant share of this growth. Although other organic sectors are growing more quickly, especially the dairy and meat sectors, the fresh produce sector remains a key ingredient to the success of the organic industry as a whole.

Over the last decade, price premiums for organic products (or the price difference between organic and comparable conventional products) have contributed to growth in certified organic farmland and, ultimately, market expansion in an industry that was formerly supply constrained. Part of the price premium may result from differential production costs (see box, "Costs of Converting to and Producing Organic Products" on page 6). Another part of the premium results from the relative level of supply and demand of organic products, which contributes to higher profits for organic farmers. Producers hope to maintain this portion of the price premium for their products as the market grows.[1] Laws of supply and demand, however, make it unlikely that price premiums contributing to higher profits and market growth can coexist over the long run: as long as higher profits exist, new suppliers will enter the market, and once market supply increases faster than demand, price premiums and the commensurate level of higher profits are likely to decline. For example, organic dairy markets in some EU countries experienced near-zero price premiums for organic milk after a significant increase in supply (*Organic Monitor*, 2002). Analysis of trends in price premiums can provide insight into relative changes in supply and demand of organic products and a clearer sense of market maturity and the likelihood of further growth rates. Studies of price premiums are limited by a lack of consistent and comparable price data.

National Organic Standards and Certification

Organic farming systems rely on ecologically based practices, such as biological pest management and composting; virtually exclude the use of synthetic chemicals, antibiotics, and hormones in crop production; and prohibit the use of antibiotics and hormones in livestock production. Under organic farming systems, the fundamental components and natural processes of ecosystems—such as soil organism activities, nutrient cycling, and species distribution and competition—are used as farm management tools.

Private organizations, mostly nonprofit, began developing certification standards in the early 1970s as a way to support organic farming and prevent consumer fraud. Some States began offering organic certification services in the late 1980s for similar reasons. The resulting patchwork of standards in the various certification programs, however, caused a variety of market problems.

Congress passed the Organic Foods Production Act of 1990 to establish national standards for organically produced commodities, and USDA promulgated final rules for implementing this legislation in December 2000, with an 18-month transition period. As of October 2002, all agricultural products that are sold, labeled, or represented as organic are to be in compliance with the regulations. They require that organic growers and handlers (including food processors and distributors) be certified by State or private agencies/organizations under the uniform standards developed by USDA, unless the farmers and handlers sell less than $5,000 a year in organic agricultural products. Retail food establishments that sell organically produced agricultural products, but do not process them, are also exempt from certification.

The national organic standards address the methods, practices, and substances used in producing and handling crops, livestock, and processed agricultural products. Although specific practices and materials used by organic operations may vary, the standards require every aspect of organic production and handling to comply with the provisions of the Organic Foods Production Act. Organically produced food cannot be produced using genetic engineering and other excluded methods, sewage sludge, or ionizing radiation. These standards include a national list of approved synthetic, and prohibited nonsynthetic, substances for use in organic production and handling.

USDA organic standards for food handlers require that all nonagricultural ingredients, whether synthetic or nonsyn-thetic, be included on the national list. Handlers must prevent the commingling of organic with nonorganic products and protect organic products from contact with prohibited substances. In a processed product labeled as "organic," all agricultural ingredients must be organically produced unless the ingredient(s) is (are) not commercially available in organic form. The labeling requirements under the national standards apply to raw, fresh, and processed products that contain organic ingredients and are based on the percentage of organic ingredients in a product. Agricultural products labeled "100-percent organic" must contain (excluding water and salt) only organically produced ingredients. Products labeled "organic" must consist of at least 95-percent organically produced ingredients. Products labeled "made with organic ingredients" must contain at least 70-percent organic ingredients. Products with less than 70-percent organic ingredients cannot use the term organic anywhere on the principal display panel but may identify the specific ingredients that are organically produced on the ingredients statement on the information panel. The USDA organic seal—the words "USDA organic" inside a circle—may be used on agricultural products that are "100-percent organic" or "organic."A civil penalty of up to $10,000 per violation can be levied on any person who knowingly sells or labels as organic a product that is not produced and handled in accordance with the regulations.

For further information, visit USDA's Agricultural Marketing Service/National Organic Program website, at www.ams.usda.gov/nop/.

MARKET FOR ORGANIC FRUITS AND VEGETABLES

Over the last decade, the U.S. organic foods industry has grown considerably. The *Nutrition Business Journal* (NBJ) estimates U.S. sales of organic foods at nearly $10.4 billion in 2003, or about 1.8 percent of total U.S. retail sales of food, up from $3.5 billion in 1997 (NBJ, 2004). The *Natural Food Merchandiser* (NFM) estimates total U.S. retail sales of organic foods at $3.3 billion in 1996, up from $1 billion in 1990 (Dimitri and Greene, 2002). Annual growth rates over the period are around 20 percent (table 1). Growth rate estimates through 2010 range from 9-16 percent annually, with growth slowing in the later years. By 2010, sales of organic foods are estimated to rise to $23.8 billion, or 3.5 percent of total retail food sales in the United States (NBJ, 2004).

Table 1. Consumer sales and growth rates of organic foods,1997-2003

Year	Sales *Bil.dols.*	Growth rate *Percent*
1997	3.6	
1998	4.3	19.7
1999	5.0	18.2
2000	6.1	21.0
2001	7.4	20.7
2002	8.6	17.3
2003	10.4	20.2

Source: Nutrition Business Journal, 2004

NBJ reports $4.3 billion in U.S. sales of organic produce in 2003, or 42 percent of total U.S. sales of organic foods. Of this amount, fresh produce accounted for the largest share (93 percent). Frozen, canned, and dried organic produce each accounted for shares of 3 percent or less (fig. 1). NBJ estimates that sales of organic fruits and vegetables in 2010 could reach $8.5 billion, an increase of more than 300 percent over sales in 2000.

Although organic food sales make up a small portion (1.8 percent) of total food retail sales in the United States, some organic fruit and vegetable categories have higher market penetration rates than others. For example, in 2002, organic fresh fruit and vegetable sales accounted for 4.5 percent of total fresh fruit and vegetable sales. Sales of organic dried beans accounted for 3.5 percent of the dried bean market. Overall category shares of organic frozen vegetables (1.8 percent), canned fruits (1.0 percent), and canned vegetables (0.6 percent) fell below the overall market share of organic food sales (NBJ, 2003).

The top fresh organic fruits and vegetables purchased in the United States are tomatoes, leafy vegetables, carrots, apples, potatoes, peaches, bananas, and squash (*The Packer*, 2002). Other high-selling produce items include strawberries, beans, mushrooms, cantaloupe, celery, broccoli, and oranges. Consumers tend to buy more organic vegetables than fruit.

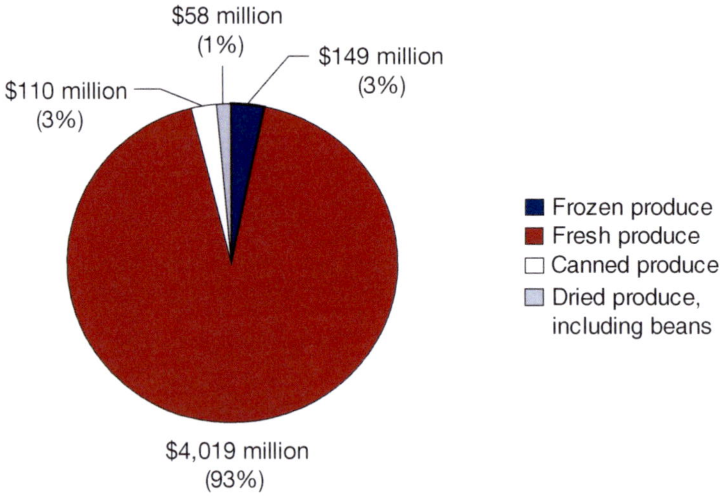

Figure 1. Sales and shares of U.S. organic produce by category, 2003

Organic produce has long been the top-selling organic food category, and it is becoming more accessible to consumers as supermarkets and other conventional retail channels continue to add the products to their shelves. In 2003, 47 percent of organic foods were sold through conventional channels, 44 percent were sold through natural food stores, and 9 percent were sold through direct and other marketing channels (e.g., farmers'markets, restaurants, exports) (OTA, 2004). The organic industry is also contributing to the fast-growing trend of packaged produce items in supermarkets (Dimitri et al., 2003). *Natural Foods Merchandiser* reports that sales of packaged fresh produce had the highest growth rate among sales of all organic products during 2002-2003, expanding 26 percent to $364 million. Conventional supermarkets accounted for three-fourths of this total. The number of new organic produce items introduced in retail markets has doubled over the last decade, from 14 in 1993 to 30 in 2003 (USDA, ERS, 2005).

Branding of fresh produce is becoming more common, much of it due to the introduction of new packaged and fresh-cut products. This trend is apparent in the organic sector, where more growers and distributors are identifying their products with their farm name and logo. Among the most prominent brands are Cal Organic, PureVeg, Earthbound Farms, and Pavich—all of which are California based, and some of which have joined forces (either through mergers or joint ventures) with conventional produce firms (NBJ, 2003).

The exact value of U.S. imports of organic fresh fruits and vegetables is unknown.[2] However, USDA's Foreign Agricultural Service (FAS) estimates the

United States imported between $1 billion and $1.5 billion in organic foods in 2002 (USDA, FAS, 2005). Imported organic fruits and vegetables must comply with U.S. national organic standards, as well as other import standards, and only certifiers or governments recognized by USDA (including international certifiers) are allowed to classify fresh produce as organic for the U.S. market. International suppliers who can provide tropical produce, off-season fresh produce (e.g., during the winter months), or in-season produce in times of domestic shortages are likely to find the greatest reception in the U.S. market. However, imported products are not assured of acceptance in U.S. markets. Recent nationwide surveys of U.S. consumers show that 41 percent of consumers purchasing organic foods seek "in-season" fresh produce, and 14 percent are concerned with country-of-origin issues (FAO/ITC/CTA, 2001).

Top 8 fresh organic fruits and vegetables purchased

1.	Tomatoes
2.	Carrots
3.	Peaches
4.	Squash
5.	Leafy vegetables
6.	Apples
7.	Potatoes
8.	Bananas

Source: *The Packer*, 2002.

U.S. exports of organic foods are also difficult to estimate. Sales of organic foods are growing worldwide, led by the EU, U.S., and Japanese markets. The global organic market was valued at $23 billion in 2002 (Willer and Yussefi, 2004). An Organic Trade Association (OTA) report (Fuchshofen and Fuchshofen, 2000) estimates U.S. organic exports in 2000 at $200 million to $300 million, and FAS estimates organic food exports in 2002 between $125 million and $250 million (USDA, FAS, 2005). The largest U.S. export categories for organic foods are soybeans, fresh and dried fruit, frozen vegetables, fruit juices, and minor food ingredients. The biggest export market for U.S. organic foods is Canada, estimated at $75 million to $150 million per year. Other major markets for U.S. organic food exports over the last 5 years include Japan, the EU, Taiwan, South Korea, New Zealand, and Australia.

Costs of Converting to and Producing Organic Products

The transition from conventional chemical-intensive farming systems to organic systems typically involves high managerial costs and risks of shifting to a new way of farming, and production costs may continue to be higher in ongoing organic operations. Costs of production can be higher in organic production systems because of a number of factors, including the relatively intensive use of labor; use of specialized equipment and other substitutes for synthetic chemicals; and high prices charged for organic seeds and other inputs. Another major factor is the use of longer crop rotations for pest and disease suppression in organic production systems. Cumulative net returns can be lower over time because high-value crops are included less frequently in organic systems than conventional systems (Temple, 2000).

The fees charged by State and private certifiers represent an additional, ongoing expense in certified organic farming systems, which can be a financial burden for smaller farmers. Certification agencies require documentation of a 3-year transition (conversion) period, during which land must be managed under approved practices, before certifying any crop or pasture acreage. Farmers cannot obtain the organic price premiums for certified organic commodities during this period, though in some cases higher prices can be obtained for "transitional" commodities. Production cost budgets are available for a number of organic fruit and vegetable production systems (Born, 2004), including small mixed-vegetable operations that are particularly common in the organic sector (Grubinger, 1999; Estes et al., 2003), as well as larger scale operations in California (University of California).

Organic produce receives a premium at all levels of the marketing chain reflecting these differences in production costs. In addition, marketing costs may also be higher for organic products because of additional processing, transportation, and handling charges. USDA organic standards require product segregation throughout the marketing chain. Also, commodities may be scarce when they are out of season in California, where domestic organic fruit and vegetable production is concentrated, and organic price premiums may be particularly vulnerable to weather conditions because of this geographic concentration (Ro and Frechette, 2001).

Organic Food Consumers

A number of industry and academic studies have examined consumer behavior to identify factors that influence purchases of organic foods. Several industry groups surveyed consumers about their preferences and buying habits for organic food. *Nutrition Business Journal* (2003) developed a consumer model based on usage and expenditure surveys performed by six agencies or companies, and reconciled it against retail and wholesale data. Based on model results for 2002, only a very small percentage of the U.S. population—2.4 percent, or 5.4 million consumers—are either frequent consumers (spending $50 a month) or heavy consumers (spending $200 a month) of organic food. A Whole Foods Market survey (2004) indicated that 54 percent of U.S. consumers have tried organic foods, and that 14 percent of the U.S. population consumed more organic food than in the prior year. The Whole Foods survey also reveals that nearly 1 in 10 Americans consume organic products regularly (several times per week).

According to Hartman's nomenclature, of the 66 percent of all consumers who claim to use organic products, 21 percent make up "core" consumers, 66 percent constitute the "midlevel," and the remaining 13 percent represent "periphery" consumers (Barry, 2004). Results of the Hartman Group's 1999 survey suggest that 3 percent of all consumers are "heavy" buyers of organic products, and 29 percent are "light" buyers (Hartman Group, 2002). The survey finds that organic vegetables and fruits fall into the top three categories of purchased organic products for both light and heavy consumers (vegetables rated second and fruits third for heavy users, and first and second for light users).

Consumer surveys also identified the factors that influence organic food purchases. In the 1980s and 1990s, concern for the environment drove consumers to purchase organic foods. However, today's consumers purchase organic foods for a variety of reasons. Respondents to a nationwide survey cited health and nutrition (66 percent), taste (38 percent), food safety (30 percent), and the environment (26 percent) as motivating factors behind organic food purchases (Hartman Group, 2002). In another survey, consumers cited the environment (58 percent), health (54 percent), food quality (42 percent), and support for small and local farmers (57 percent) as influences (Whole Foods, 2004). In addition, surveys show that consumers of organic foods are diverse in income level and race/ethnicity. Over half of those who frequently buy organic food in the United States have incomes below $30,000,[3] and African-Americans,Asian-Americans, and Hispanics use more organic products than Caucasians (Hartman Group, 2002). Price leads the list of barriers to purchasing organic products (Walnut

Acres, 2001; Whole Foods, 2004; Hartman Group, 2002), and availability of organic products is the second most often cited barrier (Hartman Group, 2002).

Academic studies have taken a slightly different tack in examining consumer behavior by complementing surveys with statistical analysis to depict the typical consumer of organic foods. According to Estes and Smith (1996), consumers consider the following factors important when purchasing fresh produce: price, size, and packaging, whether the item is on sale, and whether the item is organic. Appearance of fresh produce also matters to consumers, with the number of cosmetic defects affecting the likelihood of an organic product's being purchased (Estes and Smith, 1996; Thompson and Kidwell, 1998). A consumer's age, sex, and education level has little effect on his or her decision to buy organic produce. In fact, consumers with advanced degrees are less likely to buy organic produce (Thompson and Kidwell, 1998). Consumers concerned about food safety, as well as those who enjoy trying new products, are more likely to buy organic produce (Govindasamy and Italia, 1990). Finally, households with children under age 18 are more likely to purchase organic produce (Thompson and Kidwell, 1998).

ORGANIC FRUIT AND VEGETABLE PRODUCTION

Overall, fruit and vegetable farms made up just 6 percent of U.S. certified organic acreage in 2001, the most recent year such data are available (Greene and Kremen, 2003). In 2001, approximately 71,600 vegetable acres were certified organic, accounting for 1.6 percent of the total U.S. vegetable acreage in 2001. Approximately a third of the organic certified vegetable acreage was planted to lettuce, tomatoes, or carrots, and the rest was for other mixed vegetable acreage, other vegetable crops, and vegetable acreage that could not be classified. Nearly 5 percent of the total U.S. lettuce acreage was managed under certified organic farming systems in 2001, and nearly 4 percent of carrot acreage and 1 percent of tomato acreage was certified organic.

Certified organic fruit and tree nut acreage totaled about 55,675 acres (49,790 acres in fruits and 5,883 acres in tree nuts), accounting for approximately 1.3 percent of total U.S. fruit and tree nut acreage (Greene and Kremen, 2003). Organic grapes accounted for 29 percent of the acreage certified, followed by apples (24 percent) and citrus (20 percent); 27 percent of the total was unclassified. The unclassified category included a substantial amount of berries and stone fruits.

Among all States, California (with 40,632 acres) was the biggest organic vegetable producer in 2001, accounting for 41 percent of U.S. certified organic vegetable acreage, and the top producer of organic tomatoes, lettuce, and carrots. California also led all States in organic fruit acreage, with almost 25,000 acres, approximately half of the U.S. total. Being the top producer of organic apples,Washington had the second-largest organic fruit acreage (9,242 acres), and the second-largest organic vegetable acreage (7,174 acres). Colorado had the third-largest amount of organic vegetable acreage (4,889). Florida had the third-largest organic fruit acreage (6,154 acres) and was the top producer in organic citrus crops.

These numbers understate the importance of organic agriculture in some States. For example, certified organic acreage accounted for over 10 percent of the vegetable acreage in Vermont, New Hampshire, Maine, and Colorado in 2001, and over 2 percent of the vegetable acreage in Connecticut, Arkansas, Massachusetts, Utah,Washington, California, Oregon, and Pennsylvania. In many States, particularly in the Northeast and Southeast, most certified organic operations are small-scale farms that produce a vast array of vegetable crops, fruits, herbs, and flowers for marketing directly to consumers in the local area.

Since 1993, the Organic Farming Research Foundation (OFRF) has conducted periodic nationwide surveys of certified organic farmers. These surveys represent an effort to assess organic farmers'research and information needs, and to collect general demographic data about organic farmers across the United States. OFRF's most recent survey was released in 2004 and represented information from 2001. It was the first to focus on marketing information (Walz, 2004).

Of the almost 1,200 respondents to OFRF's survey, 43 percent produced organic vegetables in 2001, representing 9,022 acres of land in vegetable crops. In terms of sales of organic vegetables, respondents sold 74 percent as fresh market products, 19 percent as products to be processed, 3 percent as value-added products, and 3 percent as seed or propagation stock. The top four vegetables represented in the survey by acreage were spinach, sweet corn, lettuce, and broccoli.

OFRF incorporated nuts and tree crops into its fruit category, with 36 percent of respondents producing organic fruits, nuts, or tree crops in 2001, representing 6,611 acres. Farmers sold these products as fresh (55 percent), as products for processing (26 percent), as value-added products (19 percent), and as seed or propagation stock (less than 1 percent). The top three fruits represented in the survey by acreage were wine grapes, apples, and strawberries.

Organic fruit and vegetable growers made use of the same market channels, but the survey responses suggest vegetable growers relied on a wider variety of market outlets. Eighty percent of organic vegetable producers (as well as producers of herb, floriculture, mushroom, and honey products) sold their products through consumer-direct channels (e.g., onfarm sales, farmers' markets, community-supported agriculture), 69 percent sold through wholesale markets (including processors and packers, distributors, cooperatives, and supermarket buyers), and 54 percent sold through direct-to-retail markets (e.g., natural food stores, restaurants, and conventional supermarkets). In contrast, 58 percent of fruit and tree nut producers sold their products through consumer-direct channels, 50 percent sold through wholesale markets, and 38 percent sold through direct-to-retail markets.

Organic Fruit and Vegetable Prices

Many organic industry participants and observers believe that price premiums for organic products need to decrease if organic foods are to penetrate much beyond the 2- to 3-percent level into the mainstream. Some experts point to evidence that shows prices and premiums are already declining somewhat: according to interviews with growers and distributors, price volatility is still evident in organic produce but the fluctuations had evened out somewhat and organic prices are closer to conventional prices than generally perceived (NBJ, 2003). These observations are supported by economic theory, which suggests high prices and high profits accrue to the innovators and adopters of new technologies (organic, in this case), and that the profitable niche will attract new suppliers, with prices and profitability of the niche product falling over time (Blank and Thompson, 2004).

Organic farmers, however, would much rather see their products continue to command price premiums, which generate higher profits. In OFRF's 2001 survey of organic farmers (Walz, 2004), 41 percent of respondents report receiving price premiums on all of their products sold, and 71 percent received a price premium on at least half of their product sold. When asked about the circumstances that made it difficult to receive price premiums, vegetable and fruit producers cited oversupply issues and cheap imports most often, as well as limited local demand for organic products in some areas (e.g., rural areas) and price competition from conventional items (e.g., corn and strawberries). On the other hand, a number of respondents found conventional prices high enough to warrant selling their organic products at these prices.

Systematic collection of price data for organic products has been limited, thus preventing in-depth analysis of market trends in prices, margins, and price premiums for organic foods, particularly as they compare with trends for conventional foods. Though covering different years and products, several U.S. studies have examined organic price premiums. These studies of farm-level, wholesale, and retail organic price data, collected by private and nonprofit organizations, show significant premiums for organic fruits, vegetables, grains, and milk in the 1990s and beyond.

USDA tracked wholesale organic price premiums for two fresh vegetables between 1989 and 1992 and found annual average prices that were generally double conventional prices, with wide variation on a weekly basis (USDA, ERS, 2003). Monthly farmgate price premiums for several major organic fruits and vegetables consistently exceeded 100 percent between 1992 and 1996, based on reports from *Organic Food Business News* (Vandeman, 1998). Supermarket scanner data showed similar price premiums for frozen organic vegetables during this period (Glaser et al., 1998). Greene and Calvin (1997) found an average price premium of 14 percent for organic mesclun mix and 110 percent for organic carrots for a 14-month period during 1996-97. Finally, over a 17-month period during 2000-01, Sok and Glaser (2001) found an average price premium of 130 percent for organic broccoli, 125 percent for organic carrots, and 10 percent for organic mesclun lettuce at the wholesale level.

Limited data on wholesale and farmgate prices for organic products are available from USDA and private sources. USDA's Market News Service (produced by the Agricultural Marketing Service (AMS)), includes wholesale prices for organic items when available in its daily wholesale fruit and vegetable reports, which cover terminal markets in 15 U.S. cities. Organic produce prices first appeared in the *Boston Wholesale Fruit and Vegetable Report* and the *Philadelphia Wholesale Fruit and Vegetable Report* in 1992. Since then, the Market News Service has reported organic prices in a number of other wholesale markets on a sporadic basis. The Boston and San Francisco markets are the only markets, however, in which organic prices are regularly reported for approximately 10 fruits and vegetables. In a recent inventory of daily organic prices by AMS (May 2003 to May 2004), a large number of organic prices were reported for Dallas (653) and Baltimore (459), but these numbers are still much lower than for Boston (4,109) and San Francisco (1,760). The inventory also showed that prices for 58 organic fruit and vegetable commodities were reported during the same period, but only 11 had over 200 different daily price reports.

At the farmgate level, *Organic Food Business News* (OFBN) provides organic price data for fruits and vegetables gathered from a weekly nationwide survey of farmers, buyers, traders, and shippers. OFBN's data provide regular information on price levels for a variety of organic crops, over a long period of time. However, the data are from a private source and are hard to evaluate for consistency and accuracy. Comparisons of organic prices to conventional prices at the farmgate level are further limited to those items that are reported by USDA's National Agricultural Statistics Service (NASS).

Prices for Organic Broccoli, Carrots, and Mesclun Mix

Broccoli, carrots, and mesclun mix are the most consistently reported organic commodities. This report details price premiums for broccoli and carrots, both at the wholesale and farmgate level, and mesclun mix, at the wholesale level, for 2000-04. The analysis of wholesale prices is based on organic and conventional pricing data gathered by USDA's AMS. AMS reports high and low prices when available. The monthly prices used here are simple averages of the high and low prices reported.[4] Farm-level analysis is based on OFBN price data used to calculate monthly prices for organic products. OFBN provides weekly price ranges (lows and highs), and midpoints were computed on these ranges. The data provided by OFBN include a simple average of the month's midpoints. Conventional farm prices are reported as monthly averages in ERS Outlook Reports, based on monthly prices reported by NASS.

Streff and Dobbs (2004) note, however, OFBN prices may not accurately represent the prices received by most farmers because the prices are not weighted on the basis of quantities sold, but are simple highs and lows. Prices received for organic produce by different farmers can vary widely within any given month or year. While this effect may also apply to prices received by farmers producing conventional foods, price variation is likely to be greater for organic farmers. The limitation posed by averaging simple highs and lows also holds true for the wholesale-level data, for both conventional and organic prices.

Over the period 2000-04, organic price premiums were higher for both broccoli and carrots at the wholesale level than at the farmgate level (table 2). Market margins for farmgate to wholesale were higher in the organic sector than in the conventional sector for broccoli and carrots, which is consistent with earlier studies indicating that organic wholesalers earn higher margins than conventional wholesalers (Dimitri and Richman, 2000). Also consistent with earlier studies (Sok and Glaser, 2001), annual organic price premiums for broccoli and carrots are close to 100 percent or above (the highest was 180 percent in 2003 for wholesale broccoli) for all but organic farmgate carrots. Annual organic price premiums for mesclun mix, at the wholesale level, range from 6 to 9 percent over the 5-year period.

Table 2. Annual wholesale and farmgate price premiums and market margins for produce,2000-04

Item/ calendar year	Organic price premium		Market margins for farmgate to wholesale	
	Wholesale	Farmgate	Organic	Conventional
		Percent		
Broccoli				
2000	124	99	72	53
2001	141	100	93	56
2002	176	130	93	51
2003	180	133	82	42
2004	153	117	96	64
Carrots				
2000	126	117	90	71
2001	143	108	69	36
2002	152	89	64	24
2003	162	91	69	23
2004	148	75	68	22
Mesclun mix				
2000	6	n/a	n/a	n/a
2001	9	n/a	n/a	n/a
2002	8	n/a	n/a	n/a
2003	6	n/a	n/a	n/a
2004	7	n/a	n/a	n/a

Notes: n/a = not available.Organic price premiums are the percent increase over conventional prices and are calculated by subtracting the conventional price from the organic price and dividing the difference by the conventional price.Market margins are the difference between farmgate and wholesale prices.Annual price premiums and marketing margins are calculated using an average of monthly data.

Sources: Calculated by USDA's Economic Research Service from *Organic Food Business News*, 2004;USDA, ERS, 2004a-b;USDA, AMS, 2004.

Figures 2-6 detail quarterly price premiums for each commodity.[5] While quarterly prices for wholesale conventional broccoli have remained steady at mostly $8-$12 per carton (carton of 14-18 count bunches) over the last 5 years, per carton prices for wholesale organic broccoli have shown more variability, ranging from $18 to almost $38 (fig. 2). Except for two quarters, price premiums have consistently remained above 100 percent over the 5-year period, with 65 percent being the lowest price premium and 223 percent the highest.

Farmgate broccoli prices follow a similar pattern, with most conventional prices in the range of $5-$7 per 23-pound carton (comparable to the wholesale carton), with $5 the lowest and $10 the highest (fig. 3). Per carton prices for organic broccoli varied between $11 and $23. However, farmgate price premiums have been lower than those at the wholesale level. In 8 of the 20 quarters, price premiums were below 100 percent (lowest 35 percent), and the highest price premium in the 5-year period was 195 percent.

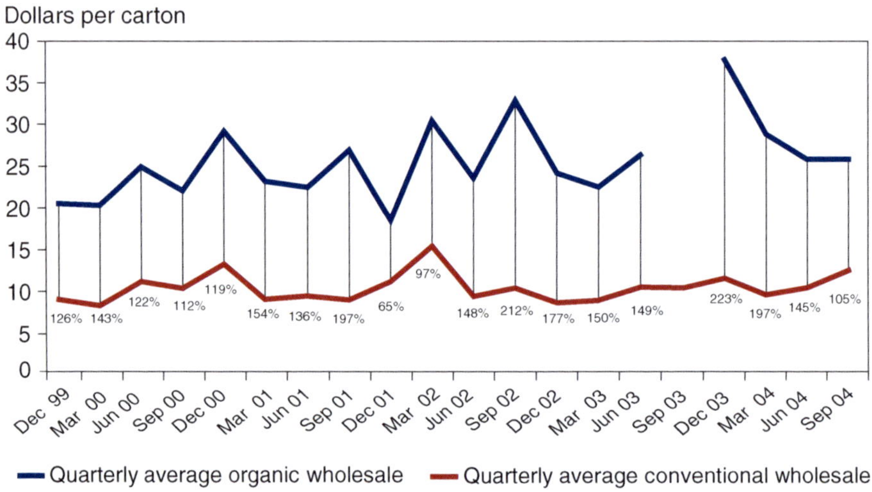

Notes: Carton includes 14-18 count bunches (23 pounds). Price premiums shown as percentages in figure. Price data not available for organic broccoli for the quarter July-September 03. (noted as Sep 03 in chart)

Source: Calculated by USDA's Economic Research Service from USDA, AMS, 2004.

See underlying data at www.ers.usda.gov/data/OrganicPrices

Figure 2. Quarterly price premiums for wholesale organic broccoli

Both quarterly wholesale conventional and organic prices for carrots have remained steady since mid-2001, leading to stable price premiums between 130 and 175 percent from third quarter 2001 to 2004 (fig. 4). Prior to 2001, organic carrot prices showed greater variability, with wholesale prices ranging from $19 per sack (sack of 24-count 2-pound bags, size medium to large) in mid-2000 to $34 per sack in mid-2001. Prices per sack steadied after 2001 and ranged from a low of $29 per sack to $32 per sack through late 2004. Conventional prices per sack have remained relatively steady throughout the 5-year period, with prices ranging from $10 to $13.

Farmgate prices for organic carrots fluctuated prior to the third quarter 2001 (ranging from $10 in mid-2000 to $20 in mid-2001) (fig. 5). Conventional prices for carrots also varied before 2001, leading to swings in price premiums in 2000 from 44 to 202 percent. Per sack prices for organic carrots at the farm level then stabilized from mid-2001 and have remained at $18.[6] After the second quarter 2001, conventional prices per sack also stabilized, ranging from $9 to $12, with organic premiums ranging from 55 to 132 percent. However, prices of conventional carrots increased in the first two quarters of 2004 and then decreased sharply in the third quarter. The relative stability of organic and conventional carrot prices likely stems from the fact that both products are grown almost exclusively under contract.

Mesclun mix showed a markedly lower wholesale price premium than the other commodities studied, ranging from -10 percent to 18 percent quarterly over the last 5 years (fig. 6). Also, unlike the other commodities whose prices have varied over the years, both organic and conventional mesclun mix have noticeably decreased in price since the late 1990s. Organic prices have fallen from around $9-$10 per carton (3-pound carton) in the late 1990s to slightly more than $6 per carton in 2004. In the conventional sector, per carton prices have fallen from $8-$9 in the late 1990s to almost $6 in 2004.[7] The price premium for organic mesclun has always been much lower than for other commodities (Greene and Calvin, 1997). Mesclun, first introduced as an organic crop in the 1990s, initially sold for high prices that attracted both organic and conventional producers to the market. As the supply of organic and conventional mesclun increased, the prices of both declined, although organic products maintain a small premium.

The trends in price premiums for broccoli and carrots over time suggest that, for the time being, even though certified organic acreage is rising rapidly, demand appears to be growing fast enough so that farmers and wholesalers are maintaining an organic premium. Interestingly, the wholesale organic premium consistently exceeds the organic premium at the farm-gate for broccoli and carrots. In contrast, wholesale price premiums and prices for both organic and conventional mesclun mix have continued to narrow since the mid-to-late 1990s.

These results cannot be generalized to the wide range of organic fruits and vegetables, and the limited availability of price data means that a broader survey of price premiums for organic goods is not possible at this time. The lack of a system for tracking organic prices at the farm, wholesale, and retail level means that—for the immediate future at least—price and premium trend analysis will be limited to these few commodities.

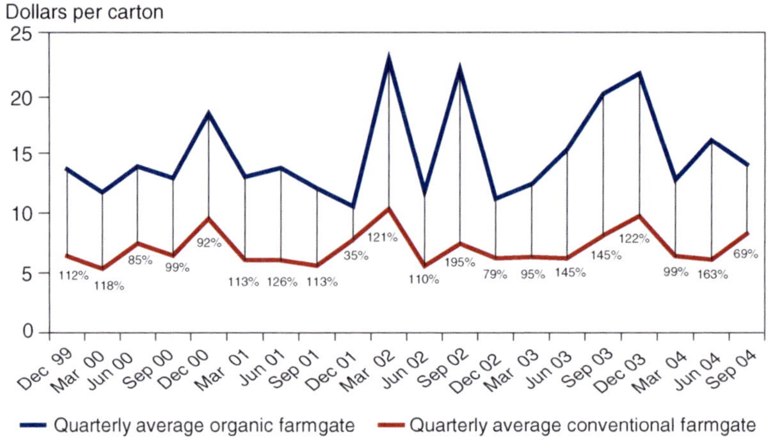

Notes: *Carton includes 14-18 count bunches (23 pounds). Price premiums shown as percentages in figure.*

Source: *Calculated by USDA's Economic Research Service from Organic Food Business News, 2004 (organic) and USDA, ERS, 2004a-b (conventional). See underlying data at www.ers.usda.gov/data/OrganicPrices*

Figure 3. Quarterly price premiums for farmgate organic broccoli

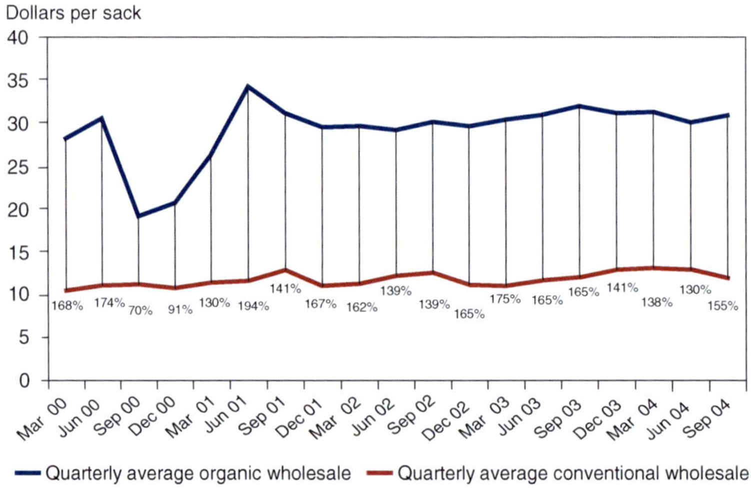

Notes: Sack includes 24-count 2-pound bags (size: medium to large). Price premiums shown as percentages in figure.

Source: Calculated by USDA's Economic Research Service from USDA, AMS, 2004. See underlying data at www.ers.usda.gov/data/OrganicPrices

Figure 4. Quarterly price premiums for wholesale organic carrots

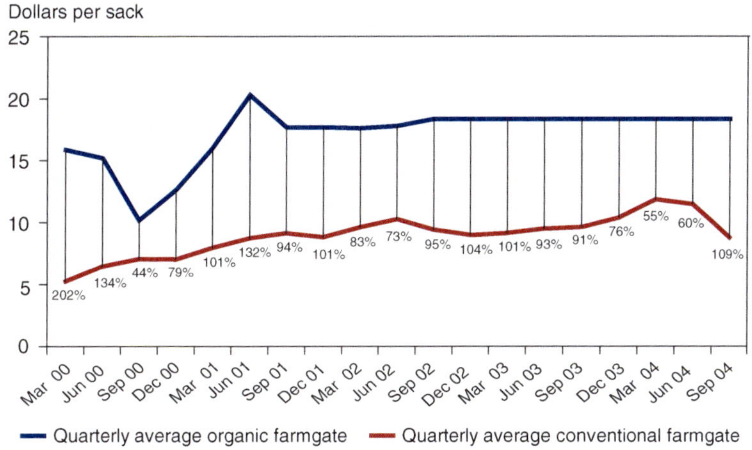

Notes: Sack includes 24-count 2-pound bags (size: medium to large). Price premiums shown as percentages in figure.

Source: Calculated by USDA's Economic Research Service from Organic Food Business News, 2004 (organic) and USDA, ERS, 2004a-b (conventional). See underlying data at www.ers.usda.gov/data/OrganicPrices

Figure 5. Quarterly price premiums for farmgate organic carrots

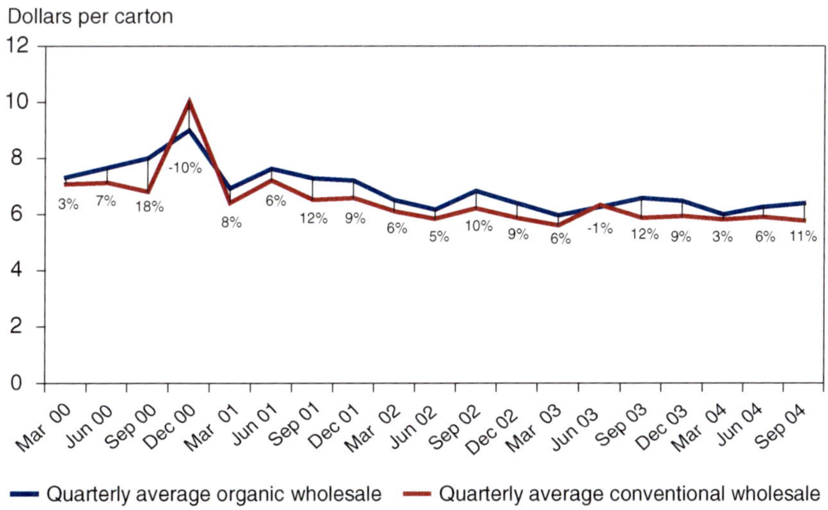

Notes: Carton includes 3 pounds of mesclun mix. Price premiums shown as percentages in figure.

Source: Calculated by USDA's Economic Research Service from USDA, AMS, 2004. See underlying data at www.ers.usda.gov/data/OrganicPrices

Figure 6. Quarterly price premiums for wholesale organic mesclun mix

LOOKING AHEAD

Fresh fruits and vegetables have long been an important component of the organic food sector, and this status is likely to continue into the future. Despite higher prices for organic products than for conventional products, the number of consumers who purchase organic produce is growing, and this group is becoming more diverse. Consumers cite various factors as influences in their decisions to purchase organically grown food, including health, nutrition, and support for the environment and for small and local farmers. For this group of consumers, rising expenditures over the last 20 years indicate that these factors outweigh the higher direct costs of organic foods. Yet, a larger body of consumers has indicated that the size of the organic price premium deters them from buying organically grown food, and thus likely will constrain market expansion at some point.

Farmers benefit by receiving higher prices, and presumably higher profits, for their organic products. A natural tension results from this effect. Higher prices and profitability encourage farmers to increase production or to enter the organic sector. If supply begins growing faster than demand, price premiums and profitability will decline. At the same time, as the price differential between organically and conventionally grown products diminishes, more consumers are likely to purchase organic food. Relative changes of supply and demand will help determine whether price premiums and higher profitability will continue for organic farmers and businesses.

For more information on U.S. organic agriculture, see ERS's organic farming and marketing briefing room, www.ers.usda.gov/briefing/organic.

BIBLIOGRAPHY

Barry, M. 2004. "The New World Order: Organic Consumer Lifestyle Segmentation." *[N]sight*. Hartman Group Fall.

Blank, Steven C., and Gary D. Thompson. 2004. "Can/Should/Will a Niche Become the Norm? Organic Agriculture's Short Past and Long Future." *Contemporary Economic Policy*, 22(4):483-503.

Blisard, N., H. Stewart, and D. Jolliffe. 2004. *Low-Income Households' Expenditures on Fruits and Vegetables*,Agricultural Information Bulletin No. 792-5. U.S. Department of Agriculture, Economic Research Service, October, www.ers.usda.gov/publications/aib792/aib792-5/aib792-5.pdf.

Born, H. 2004. *Enterprise Budgets and Production Costs for Organic Production*,ATTRA-NCAT, National Center for Appropriate Technology, www.attra.ncat.org/.

Dimitri, C.,A. Tegene, and P. Kaufman. 2003. *U.S. Fresh Produce Markets: Marketing Channels, Trade Practices, and Retail Pricing Behavior*, Agricultural Economic Report No. 825, U.S. Department of Agriculture, Economic Research Service, September, www.ers.usda.gov/Publications/ aer825/.

Dimitri, C., and C. Greene. 2002. *Recent Growth Patterns in U.S. Organic Foods Market.* Agricultural Information Bulletin No. 777, U.S. Department of Agriculture, Economic Research Service, September, www.ers.usda.gov/publications/aib777/.

Dimitri, C., and N. Richman. 2000. *Organic Food Markets in Transition*, Public Policy Report No. 14, Henry A. Wallace Institute for Alternative Agriculture,April.

Estes, E., T. Kleese, and L. Lauffer. 2003. *North Carolina Organic Vegetable Production Cost Study*, Department of Agricultural and Resource Economics, North Carolina State University and Carolina Farm Stewardship Association.

Estes, E., and V. Kerry Smith. 1996. "Price, Quality, and Pesticide-Related Health Risk Considerations in Fruit and Vegetable Purchases:An Hedonic Analysis of Tucson,Arizona Supermarkets," *Journal of Food Distribution Research* 27(3): 8-17, Fall.

Food and Agriculture Organization of the United Nations, International Trade Centre, and Technical Centre for Agricultural and Rural Cooperation (FAO/ ITC/CTA). 2001. *World Markets for Organic Fruit and Vegetables - Opportunities for Developing Countries in the Production and Export of Organic Horticultural Products*, FAO.

Fuchshofen,W., and S. Fuchshofen. 2000. *Organic Trade Association's Export Study for U.S. Organic Products to Asia and Europe*, Organic Trade Association, www.ota.com/organic/mt/export_form.html.

Glaser, L., B. Krissoff,A. Lengyel, and K. Sheehan. 1998. "Demand for Frozen Vegetables:A Comparison of Organic and Conventional Products," *Vegetables and Specialties Situation and Outlook.* VGS-276. U.S. Department of Agriculture, Economic Research Service, November.

Govindasamy, R., and J. Italia. 1990. "Predicting Willingness to Pay a Premium for Organically Grown Fresh Produce," *Journal of Food Distribution Research*,Volume 30 (2):44-53.

Greene, C., and L. Calvin. 1997. "'Organically Grown'Vegetables: U.S. Acreage and Markets Expand During the 1990s", *Vegetables and Specialties Situation and Outlook*,VGS-271, U.S. Department of Agriculture, Economic Research Service, May.

Greene, C., and A. Kremen. 2003. *U.S. Organic Farming in 2000-2001: Adoption of Certified Systems*. Agriculture Information Bulletin No. 780, U.S. Department of Agriculture, Economic Research Service,April.

Grubinger,V. 1999. *Sustainable Vegetable Production from Start-up to Market*, NRAES-104, Natural Resource,Agriculture, and Engineering Service—Cooperative Extension, Cornell University.

Hartman Group. 2002. *Hartman Organic Research Review: A Compilation of National Organic Research Conducted by the Hartman Group*. Bellevue, WA.

Hartman Group. 2004. *Organic Food and Beverage Trends, 2004*. Bellevue, WA.

Natural Foods Merchandiser, various issues.

Nutrition Business Journal (NBJ). 2003. *The NBJ/SPINS Organic Foods Report 2003*, Penton Media, Inc.

Nutrition Business Journal (NBJ), 2004. *The NBJ/SPINS Organic Foods Report 2004*, Penton Media, Inc.

Organic Food Business News. 2004. *Organic Commodity Prices*. Hotline Printing and Publishing. November 1999-November 2004.

Organic Monitor, 2002. "U.K. Future Development of Organic Dairy Sector." December 20. London, UK. www.organicmonitor.com

Organic Trade Association (OTA). 2004. *The OTA 2004 Manufacturers' Survey Overview*.

The Packer. 2002. "Fresh Trends: Profile of the Fresh Produce Consumer".

Ro, E., and D. Frechette. 2001. "The Effect of California Weather Conditions on Price Premia for Organically Grown Vegetables in the United States," *Journal of Food Distribution Research*, July.

Sok, E., and L. Glaser. 2001. "Tracking Wholesale Prices for Organic Produce," *Agricultural Outlook*, U.S. Department of Agriculture, Economic Research Service, October, www.ers.usda.gov/publications/agoutlook/oct2001/ao285d.pdf.

Streff, N., and T. Dobbs. 2004. *'Organic' and 'Conventional' Grain and Soybean Prices in the Northern Great Plains and Upper Midwest: 1995 through 2003*, Econ Pamphlet 2004-1, South Dakota State University, Economics Department, June.

Temple, S. 2000. *The Transition from Conventional to Low-Input or Organic Farming Systems: Soil Biology, Soil Chemistry, Soil Physics, Energy Utilization, Economics, and Risk*, Sustainable Agriculture Research and Education Program-Final Report, University of California, November.

Thompson, G., and J. Kidwell. 1998. "Explaining the Choice of Organic Produce: Cosmetic Defects, Prices, and Consumer Preferences," *American Journal of Agricultural Economics* v80, n2 (May): 277-8.

University of California. Various years. *Organic Farming Cost and Returns Studies*, Sustainable Agriculture Research and Education Program, University of California, www.sarep.ucdavis.edu/pubs/Costs.htm.

U.S. Department of Agriculture,Agricultural Marketing Service (USDA, AMS). 2004. *Boston Wholesale Fruit and Vegetable Report*, September 1999-December 2004.

U.S. Department of Agriculture, Economic Research Service (USDA, ERS). 2005. "New Organic Product Introductions Are Led by Beverages, Prepared Foods, and Snacks," *Amber Waves*,Vol. 3, Issue 2, p. 44,April, www.ers.usda.gov/AmberWaves/April05/Indicators/researchareas.htm/.

U.S. Department of Agriculture, Economic Research Service (USDA, ERS). 2004a. *Fruit and Tree Nuts Situation and Outlook Yearbook*, FTS-2004S, October, www.ers.usda.gov/Publications/fts/.

U.S. Department of Agriculture, Economic Research Service (USDA, ERS). 2004b. *Vegetables and Melons Situation and Outlook Yearbook*,VGS-2004, July, www.ers.usda.gov/Publications/vgs/.

U.S. Department of Agriculture, Economic Research Service (USDA, ERS). 2003. *Briefing Room—Organic Farming and Marketing: Questions and Answers*, June, www.ers.usda.gov/Briefing/Organic/Questions/orgqa6.htm/.

U.S. Department of Agriculture, Foreign Agricultural Service (USDA, FAS). 2005. *U.S. Market Profile for Organic Food Products,* February.

Vandeman,Ann. 1998. "Price Premiums for Organic Fresh and Processed Foods," In *Organic Farming and Marketing Research—New Partnerships and Priorities, Workshop Proceedings.* U.S. Department of Agriculture, Cooperative State Research, Education, and Extension Service, October.

Walnut Acres. 2001. "The Walnut Acres Certified Organic Future," www.walnutacres.com/.

Walz, Erica. 2004. *Final Results of the Fourth National Organic Farmers' Survey: Sustaining Organic Farms in a Changing Organic Marketplace*, Organic Farming Research Foundation, www.ofrf.org/publications/survey/index.html/.

Whole Foods. 2004 "Organic Foods Continue to Grow in Popularity According to Whole Foods Market Survey," October, www.wholefoods.com/company/pr_10-21-04.html/.

Willer, H., and M. Yussefi. 2004. *The World of Organic Agriculture: Statistics and Emerging Trends 2004*, International Federation of Organic Agriculture Movements, Bonn, Germany.

ACKNOWLEDGMENTS

We thank Shubha Chakravarty (formerly with ERS) for invaluable research assistance with the price data and Fred Teensma (Agricultural Marketing Service) for guiding us to various data sets. We also thank Terry Long (Agricultural Marketing Service), Jim Tringe (Foreign Agricultural Service), Ted Jaenicke (Pennsylvania State University), Rick Welsh (Clarkson University), Joy Harwood (ERS), and Lewrene Glaser (ERS) for thoughtful reviews of the manuscript. Appreciation is also extended to the editors, John Weber and Tom McDonald, and to Anne Pearl for text layout and graphic design.

REFERENCES

[1] Any part of a price premium that may be due to differential production costs will be maintained as the market grows. Additional research is needed to understand the components of the organic price premium.

[2] Measuring organic imports and exports is difficult because current codes (International Trade Data System) used to analyze trade flows of agricultural products do not separate organic products from conventionally produced products. In the future, tracking the value of products declared as "USDA Organic" may be possible through a database initiative being undertaken by multiple agencies—the Automated Commercial Environment (part of the International Trade Data System) (USDA, FAS, 2005).

[3] The results of the Hartman survey are interesting, given that a USDA ERS study that found that low-income households eat less fresh fruits and vegetables than higher-income households (Blisard et al., 2004), suggesting that additional research on who buys organic foods is necessary.

[4] The monthly price averages the month's daily average of the reported high and low prices.

[5] Quarterly prices were calculated by averaging the monthly prices in the organic farmgate and wholesale prices data product at www.ers.usda.gov/data/OrganicPrices Quarterly price premiums are calcu-

lated using quarterly organic prices and quarterly conventional prices. Organic price premiums are calculated by sub- tracting the conventional price from the organic price and dividing the differ- ence by the conventional price.

[6] Because of the proprietary nature of the data, it is not clear why organic carrot farmgate prices have remained constant at $18.35 since July 2002.

[7] Only wholesale prices for mesclun mix are available, since NASS does not collect farmgate price information on this commodity.

In: Organic Agriculture in the U.S.
Editor: Alison J. Wellson, pp. 97-127
ISBN: 978-1-60021-305-2

Chapter 4

MARKET-LED VERSUS GOVERNMENT-FACILITATED GROWTH DEVELOPMENT OF THE U.S. AND EU ORGANIC AGRICULTURAL SECTORS

Carolyn Dimitri and Lydia Oberholtzer

ABSTRACT

Organic farmland and sales are rapidly growing worldwide, and the two largest markets are in the European Union (EU) and the United States. The two regions have adopted different policy approaches to organic agriculture. Many EU countries have "green payments" available for transitioning and continuing organic farmers, as well as a variety of other supply and demand policies aimed at promoting growth of the organic sector. The U.S. Government, in contrast, has largely taken a free-market approach to the organic sector, and policy is aimed at facilitating market development. This report compares EU and U.S. organic agriculture policy and examines the organic sectors in the two regions.

Keywords: Organic agriculture, green payments, farm policy, organic markets, organic certification, EU organic sector, U.S. organic sector

INTRODUCTION

The organic sector is rapidly growing in the European Union (EU) and the United States. Together, consumer purchases in these two regions made up 95 percent of the €19 billion (€ = euro), or $25 billion, in estimated world retail sales of organic food products in 2003 (Willer and Geier, 2005).[1] Agricultural land under organic production has also increased rapidly in both regions. Government policy regarding organic agriculture differs markedly in the two regions, however. The EU actively promotes sector growth via conversion subsidies and direct payments to farmers, while the U.S. largely takes a free-market approach, with policies that focus on facilitating market development. As a result of the different policies, the EU-15 has more certified organic farmland than does the U.S. (4.4 million hectares versus 949,000 hectares in 2001), a greater share of farmland under organic management (2 percent versus 0.25 percent in 2001), and more organic farms (143,607 versus 6,949 in 2001) (see box, "EU and U.S. Organic Sectors").[2,3]

The reasons for the different policy approaches may be that the EU and U.S. governments have inherently different ideas about agriculture, the environment, and by extension, organic agriculture. From the perspective of many EU countries, organic agriculture delivers environmental, social, and other benefits to society and is an infant industry that needs support until it is mature and able to compete in established markets (Lampkin, 2003). The U.S. Government, while acknowledging organic agriculture's positive impact on soil quality and erosion, considers the organic sector primarily an expanding market opportunity for producers and regards organic food as a differentiated product available to consumers (Robinson, 2003; USDA, AMS, 2002). Despite policy differences, organic agriculture and sales are rapidly increasing in both regions. In comparison, other sectors of agriculture have been stagnating in developed countries.

Based on farmland acreage and number of farms, European farmers are able to supply more organic products than are their American counterparts, thus making international trade important to the global organic market (see figure). The United States, once a net exporter of organic products, imported an estimated $1.0-$1.5 billion in organic food in 2002, and the ratio of imported to exported organic products was about 8 to 1 (USDA, FAS, 2005). Imported products from the EU include dry pasta, processed tomatoes, wine, and olive oil. The value of U.S. organic exports to the EU was estimated at $100-$200 million a year in 1998 (USDA, FAS, 1999). Once the U.S. and EU countries agree on equivalency of standards, trade opportunities are likely to expand.

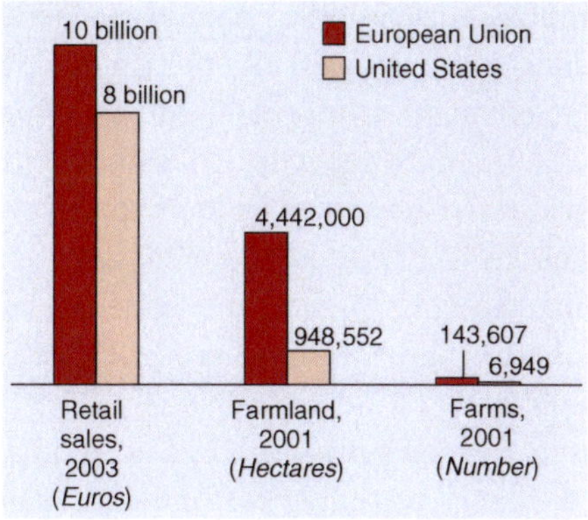

Source: Greene and Kremen, 2003; Lampkin, 2005a and b; Sahota, 2004.

EU and U.S. Organic Sectors

Despite the large presence of the organic market in both the EU and U.S., the organic industry began in Europe. Organic agriculture emerged in the 1920s, first in England (via the work of Sir Robert McCarrison, Sir Albert Howard, and Richard St. Barbe Baker) and Germany (via Rudolf Steiner) (Conford, 2001). Organic agriculture did not cross the ocean until years later, when a student of Steiner, Ehrenfried Pfieffer, moved to the U.S. in 1938. J.I. Rodale, along with others, facilitated further growth in the U.S. organic industry (Conford, 2001). In both regions, organic agriculture has made great strides since these early pioneers began farming organically.

In this report, we examine organic consumers, market growth, and land under organic production in the U.S. and the EU. We then focus on the policies developed in both regions to support organic production and markets and examine the similarities and differences. The comparison illuminates the fact that policy directions are tied to a society's point of view—in this case, about the costs and benefits of conventional and organic agriculture—and that policies, in turn, influence the trajectory of markets (both supply and demand), research agendas, and institutional development.

EU AND U.S. ORGANIC CONSUMERS

In many ways, development of the European and U.S. markets has followed a similar path. In the early days, the organic sector was supply driven and organic products were introduced by farmers; these pioneers had different reasons for raising food organically. In some cases, the farmers were seeking to reduce the use of chemicals in farming, thus restoring natural soil fertility; others adopted organic systems for spiritual reasons; and still others turned to organic farming in response to growing government involvement in agriculture (Conford, 2001; Dimitri and Richman, 2000). More recently, consumers have been the driving market force in both regions. Consumers in both parts of the world share a number of reasons for their growing interest in purchasing organic food, with food safety and health topping the list. At the same time, distinct experiences and cultural perspectives in the EU and U.S. have brought about some divergence in motives for buying organic food.

One noteworthy issue for consumers of organic food is the "food scare," which has the potential to change consumer demand for certain products and their substitutes. European consumers have faced more serious food scares than have American consumers. For example, mad cow disease (bovine spongiform encephalopathy, BSE) was an enormous boon to the European organic livestock and dairy industry, as many consumers substituted organic dairy and meat products, which they perceived as safer, for conventionally raised dairy and meat products. In 2001, organic food sales in Germany increased by 30 percent as a result of BSE (USDA, FAS, 2002). Other food scares that have shifted European consumers toward organic food products include episodes of contaminated chicken feed in Belgium in 1999, feed contaminated by dioxin in 2004, and most recently, carcinogenic food dyes in TV dinners in Ireland in 2005.[4] In contrast, U.S. consumers have not been strongly affected by food scares. For example, the Alar scare in the late 1980s caused a temporary rise in organic food sales. More recently, the two confirmed cases of mad cow disease appear to have had little impact on the organic livestock and dairy sector.

Cultural perspectives also influence demand for organic food. For example, European consumers have generally expressed opposition to biotech or genetically modified (GM) food. In contrast, U.S. consumers have been largely unaware of biotech food introductions, and objections to GM foods have not been strong. However, U.S. organic consumers vocally opposed GM foods, as indicated by the reaction to the U.S. Department of Agriculture's (USDA) draft rule that did not prohibit the use of GM ingredients in organic food. The final rule reflects this consumer desire and disallows classifying food produced with GM ingredients as organic, which is also true in the EU.

While food scares have provided bursts of energy to the organic market, other key motivating factors may be more important in sustaining the market in the long run. These factors have changed as the market has evolved. Although U.S. consumers in the 1980s and earlier purchased organic food because of their concern for the environment (Barry, 2004), environmental concerns now rank fourth according to a nationwide survey conducted by the Hartman Group (2002). The Hartman survey revealed that two-thirds of organic consumers cited health and nutrition as a reason for buying organic, followed by taste (38 percent), food safety (30 percent), and the environment (26 percent).

Studies indicate that most European consumers have made a similar shift, from purchasing organic food for altruistic reasons to more self-interested reasons, such as food safety and health. Michelsen and colleagues (1999) found that food safety and health were the most important arguments made by EU retailers in promoting their organic products, whereas environmental protection was the second most important argument. Ranking behind these two top motivators were taste, nature conservation, and animal welfare. Dabbert and colleagues (2004) suggest possible regional differences in Europe as well. For example, animal welfare and environmental issues do not seem to motivate consumers in Mediterranean countries, such as Italy and Greece, but do play an important role in Northern Europe.

Consumers in both regions face similar impediments to purchasing organic food. According to consumer surveys, price leads the list of barriers to purchasing organic products in the U.S, followed by availability of organic products (Hartman Group, 2002; Walnut Acres, 2001; Whole Foods 2004). In Europe, the main obstacles include high prices, poor product distribution, little obvious difference in quality, lack of information on the nature of organic products, and doubts about the organic integrity of the items (Dabbert et al., 2004).

EU and U.S. Market Growth

Comparing growth rates of retail sales in the two regions suggests that the European organic market is more mature than the U.S. market. The EU—particularly the Western European countries—boasted the fastest growth in the organic sector in the 1990s. Growth rates for organic retail sales, however, have slowed in some countries, with recent growth across the EU averaging 7.8 percent per year (Organic Monitor, 2003). Growth rate forecasts range from 1.5 percent for Denmark to 11 percent for the United Kingdom (UK) annually for the next few years (Willer and Richter, 2004).

Growth rate diversity across the EU is the result of the different stages of development of the organic markets in different countries. For example, the organic markets in Greece and Portugal are in the initial high-growth phase; those in Italy, France, and the UK are starting to emerge from the initial boom phase; and those in Austria, Denmark, and Sweden are more mature with slower growth rates.[5] In contrast to EU growth rates, those for organic retail sales in the U.S. averaged 20 percent per year throughout the 1990s and are predicted to be in the range of 9-16 percent through 2010 (Dimitri and Greene, 2002; Nutrition Business Journal, 2004).

In 2003, retail organic sales in the EU exceeded those in the U.S., with the EU market estimated at almost €10 billion (almost $13 billion) and the U.S. at about €8 billion (or $10.3 billion) (table 1). Germany led the European countries in organic retail sales with €3.1 billion in 2003, followed by the UK and France (about €1.6 billion each) and Italy (€1.4 billion). In 2001, Denmark had the highest market penetration of organic products (2.5-3.0 percent), followed by Austria (2 percent) and Germany (1.25-1.50 percent). Ireland had the smallest market penetration (less than 0.5 percent).

In the U.S., penetration in the food market for organic products was estimated to be 1.4 percent in 2001 (table 1) and reached 1.8 percent in 2003. The Nutrition Business Journal (2003) estimates that market penetration for organic foods will reach 3.5 percent by 2010. The organic products with the greatest penetration in the U.S. food market in 2002 include soymilk (65 percent of the market), nonsoy dairy alternatives (18 percent), and baby food, fresh juice, and produce at about 5 percent each.

Sales channels for organic products vary widely among the countries. In 2003, organic food sales in the U.S. were distributed almost evenly between natural product and health food stores (47 percent) and conventional retail stores (44 percent), with direct sales and exports accounting for 9 percent. This is a dramatic shift from 1998 when corresponding sales were 63 percent, 31 percent, and 6 percent (Nutrition Business Journal, 2003; Organic Trade Association, 2004).

As in the U.S., mainstream European supermarkets stock a wide range of organic products. However, the distribution of products through different retail outlets varies across countries (Dabbert et al., 2004). For example, over 85 percent of organic products are sold through general food shops in Denmark compared with less than 5 percent in Luxembourg. In fact, in most Scandinavian countries, the UK, and Austria, organic food is generally sold through supermarkets. In contrast, in Luxembourg and Greece, it is primarily sold through other stores (e.g., organic/health food stores, bakers, and butchers). Anumber of countries, including Ireland, Italy, France, Belgium, The Netherlands, and Germany to some extent, are more evenly divided between supermarkets and other stores.

Although the market for organic products is growing in both the EU and the U.S., there are some problems with the flow of products to the market. In Europe, the organic dairy and livestock industries, in particular, have grown rapidly over the last decade, and in some cases, more quickly than the market and distribution channels can handle. Organic milk supplies in some regions were large enough to reduce organic prices, causing some producers to exit the sector because they were unable to turn a profit (Hallam, 2003; Kortbech-Olesen, 2003). The milk glut, however, appears to be giving way to shortages in the UK, as demand continues to grow and supply has declined (Organic Monitor, 2005).

While the U.S. organic food market was formerly supply-constrained, there are indications that the market is currently better able to meet consumer demand, especially for fresh produce (Dimitri and Richman, 2000; Nutrition Business Journal, 2004). In the dairy market, however, where demand is rapidly increasing, suppliers are struggling to provide enough organic milk to satisfy demand at current prices (*Organic Business News*, 2004).

Most organic products sell for a premium over comparable conventional products, likely due in part to higher production, processing, procurement, and distribution costs, relative to those of conventional products. These higher costs may result from the wide dispersion of organic farms and their relatively small production levels, which would increase transaction costs.[6] Another part of the premium comes from the relative supply and demand levels of organic products, which contribute to higher profits for organic farmers. Price premiums to farmers vary by country and product in the EU. In 2000, meat and dairy received the lowest premiums, averaged across EU countries, with milk at 22 percent, beef, 34 percent, and pork, 68 percent; poultry and eggs averaged 182 percent and 167 percent, respectively; and grains averaged 102 percent (Schneider et al., 2005).

In 2002, the Organic Farming Research Foundation conducted its fourth biennial survey. According to the survey, 71 percent of responding U.S. organic farmers received a price premium for at least half their farm products in 2001. In addition, 45 percent sold organic products as conventional products, and half of those sold in conventional markets because the organic market was unavailable. The survey also indicated that 15 percent of organic farmers realized price declines in 2001, while the remaining 85 percent either had prices rise or remain steady over previous years (Walz, 2004). Systematic collection of price data for organic products has been limited in the U.S., thus preventing in-depth analysis of market trends in prices, margins, and price premiums between organic and conventional products. The few U.S. studies of farm-level, wholesale, and retail organic price data have shown significant organic premiums for most fruits, vegetables, grains, and milk (Bertramsen and Dobbs, 2002; Glaser et al., 1998; Glaser and Thompson, 2000; Oberholtzer, Dimitri, and Greene, 2005; Sok and Glaser, 2001; Streff and Dobbs, 2004).

Table 1. EU and U.S.organic sales

Country	Retail sales, 2003	Share of total food sales, 2001	Per capita spending on organic food, 2003
	Million	*Percent*	*Euros*[1]
Austria	400	2.0	48.9
Belgium	300	NA	NA
Denmark	339	2.5-3.0	113.6
Finland	212	NA	NA
France	1,578	1.0	21.2
Germany	3,100	1.25-1.50	30.4
Greece	21	NA	NA
Ireland	40-50	<.5	NA
Italy	1,400	1.0	19.2
Luxembourg	NA	NA	NA
The Netherlands	395	.75-1.0	38.1
Portugal	NA	NA	NA
Spain	144	NA	NA
Sweden	420	1.0	45
United Kingdom	1,607	1.0	15.3
European Union	9,966	NA	NA
United States	8,019	1.4	35.7

NA = Not available.

Note: U.S.retail sales dollars were converted to euros on an exchange rate of $1.29 = 1.00, May 2005.

[1]Per capita spending figures based on an exchange rate of $1.00 =1.00.

Sources: European sales Richter, 2005;Ireland Häring et al., 2004;U.S.sales Nutrition Business Journal, 2004, 2003;per capita expenditures Europe:Häring et al., 2004; U.S.:USDA, FAS, 2005.

EU AND U.S. ORGANIC FARMLAND

Europe has 23 percent of the world's organic farmland.[7] Certified organic land in the EU rose from 103,000 hectares (254,519 acres) in 1985 to 5.1 million hectares (12.6 million acres), or about 4 percent of the total agricultural area of 127 million hectares, in 2003 (Lampkin, 2005a). Of the 942 million acres (381 million hectares) of total U.S. agricultural lands, certified organic acreage increased 1.3 million acres (544,933 hectares) to 2.34 million (948,552 hectares) between 1997 and 2001 (Greene and Kremen, 2003). Thus, in 2001, the EU had almost five times the amount of organic farmland as did the U.S., while the U.S. had three times as much agricultural land (table 2).

Table 2. EU and U.S.organic land,1997-2003

Country	1997	1998	1999	2000	2001	2002	2003
				Hectares			
Austria	345,375	287,899	272,635	272,000	276,410	296,154	328,803
Belgium	6,818	11,744	18,515	20,667	22,452	24,000	24,000
Denmark	59,963	93,201	136,629	157,661	168,377	174,350	165,146
Finland	102,342	126,176	136,662	147,268	147,943	156,692	159,987
France	165,406	218,775	315,771	369,933	419,750	517,965	550,000
Germany	389,693	414,293	452,327	546,023	632,165	696,978	734,027
Greece	10,025	15,402	21,451	26,707	31,118	77,120	244,455
Ireland	18,687	24,411	29,360	27,231	30,017	29,754	28,514
Italy	641,149	785,738	911,068	1,040,377	1,237,640	1,168,212	1,052,002
Luxembourg	618	744	888	1,074	2,003	2,852	3,002
The Netherlands	16,960	22,268	26,350	32,334	35,876	42,610	41,865
Portugal	12,193	29,533	47,974	50,002	70,857	91,006	120,729
Spain	152,105	269,465	352,164	380,920	485,079	665,055	725,254
Sweden	118,175	127,330	155,463	174,227	202,827	214,120	225,776
United Kingdom	54,670	274,519	390,868	527,323	679,631	741,174	695,619
European Union[1]	2,094,179	2,701,498	3,268,125	3,773,747	4,442,145	4,898,042	5,099,179
United States[2]	554,933	NA	NA	821,137	948,552	NA	NA

NA = Not available.

[1]All EU hectares are for certified organic and in-conversion land. Numbers for Sweden (as well as Germany in 1997) do not reflect the substantial hectares that are managed organically but not certified.In Sweden, these lands are financially supported by the government through green payments, as recognition by Sweden and increasingly other Scandinavian countries that financially supporting organic land management for environmental gain does not necessarily need to be linked to the marketing of organic food, for which certification is a legal requirement.In Sweden, these lands accounted for another 180,000 hectares in 2003.We thank Nicolas Lampkin for pointing out the intricacies of European governmental support.

[2]The U.S. reports certified organic acreage, which has been converted to hectares (1 acre = 0.405 hectares.). Sources:EU data as of May 31, 2005, are from Lampkin, 2005a and c.U.S.data are from Greene and Kremen, 2003.

Table 3. EU and U.S.organic farm sector statistics

Country	Organic farms	Farms under organic production	Farmland under organic production
	Number	----------*Percent*----------	
Austria	19,056	9.5	9.7
Belgium	688	1.1	1.7
Denmark	3,510	6.1	6.1
Finland	5,074	6.3	7.2
France	11,377	1.7	1.9
Germany	16,476	3.5	4.3
Greece	6,028	.7	6.2
Ireland	889	.6	.7
Italy	44,039	2.0	6.9
Luxembourg	59	2.0	2.4
The Netherlands	1,522	1.5	2.2
Portugal	1,507	.4	3.2
Spain	17,028	1.3	2.8
Sweden	3,562	4.4	7.4
United Kingdom	4,017	1.7	4.4
European Union, 2003	134,434	2.0	3.9
United States, 2001	6,949	.3	.3

Note: These numbers include farms that are certified organic and in-conversion.Sweden s numbers do not include lands that are managed organically or allowed green payment support but not certified.

Sources: EU data as of June 13, 2005, are from 2003 in Lampkin, 2005a-c.U.S.data are from 2001 and detailed in Greene and Kremen, 2003.

In 2003, Italy had about one-fifth of the organic farmland in the EU, with about 1 million hectares of organic farmland, followed by Germany (734,027 hectares), Spain (725,254 hectares), and the UK (695,619 hectares) (table 2). Luxembourg accounted for the smallest amount of organic farmland, only 3,002 hectares. According to Lampkin (2005a), Austria had the highest share of farmland area under organic management, about 9.7 percent of total hectares (Austria's hectares rose significantly in 2003), followed by Sweden (7.4 percent), Finland (7.2 percent), and Italy (6.9 percent) (which lost significant organic farmland in 2003) (table 3). In contrast, in 2001, the U.S. share of farmland under organic management was 0.25 percent. States with the greatest amount of organic cropland were California, North Dakota, Minnesota, Wisconsin, Iowa, and

Montana, while Colorado and Texas had the most organic pasture and rangeland (Greene and Kremen, 2003).

The EU has far more organic farms (134,434 in 2003) than does the U.S. (6,949 in 2001) (table 3). Of the total number of EU organic farms in 2003, Italy accounted for almost one-third with 44,039 farms, followed by Austria, Spain, Germany, and France (ranging from 19,056 to 11,377 farms). Luxembourg trailed with 59 organic farms. Austria had the highest share of farms under organic production—9.5 percent (Lampkin, 2005c). In the U.S., by contrast, 0.3 percent of all farms were in organic production in 2001. The top 5 States included California, with over 1,000 operations, followed by Washington (548), Wisconsin (469), Minnesota (421), and Iowa (384) (Greene and Kremen, 2003).

EU and U.S. Organic Agriculture Policies

U.S. organic agriculture policies primarily focus on demand, through recently enacted national standards and certification, but the policies also include a limited number of federally funded grants that support research (including on-farm research), education, and marketing. In contrast, in keeping with their belief that organic farming provides benefits to society, EU countries have implemented a wide variety of policies designed to increase the amount of land farmed organically. The programs affect both supply and demand in the sector and include national standards and certification, conversion and support payments for farmers, targets for land under organic management, and other supporting policies for research, education, and marketing.

Organic agriculture policies are a subset of the EU's general agri-environmental policies, in which the framework and cofinancing are provided by the EU. Each country ultimately decides which set of policies to implement. Some member states have adopted policies that are aligned with the U.S. approach, in that they mainly pursue free-market approaches (such as The Netherlands), while many others actively promote organic agriculture. EU member states are currently debating the optimal mix of policies to support organic agriculture.[8] The general set of agri-environmental policies is complex. The following discussion highlights some of the policies; for a more complete and detailed discussion, see Dabbert et al., 2004; Häring et al., 2004; and Lampkin et al., 1999.

EU AND U.S. ORGANIC AGRICULTURE STANDARDS AND CERTIFICATION

Organic standards are set by the governments of the EU countries and the U.S. There are two economic reasons for adopting standards. Standards reduce transaction costs by providing consumers, intermediaries, and producers with a standardized definition of an organic product, thus facilitating local and international trade. Standards also resolve an information problem resulting from the unobservability of the "organic" status of a product to consumers; although the producer has full knowledge of the production methods, to the buyer, they are unobservable. Certification transmits information that the food was raised, distributed, and processed appropriately, while standards define the types of systems that are legitimately organic. An official certification system reduces opportunistic behavior (such as falsely claiming a product is organic), not only by defining the certification process and standards, but also by creating a specific enforcement system. In the U.S., for example, penalties are clearly outlined for firms that use the organic seal inappropriately, while the EU regulation does not specify enforcement mechanisms, leaving enforcement up to individual member states (Grolink, 2002). Thus, certification and standards reduce costs by facilitating information transmission and enforcing the government-sanctioned definition of "organic."

The essence of organic agriculture is ecologically farmed land and a careful eye toward soil fertility maintenance. Because the actual techniques used in ecological farming are region specific, individual countries often have idiosyncratic definitions of organic, all of which are based on defining an ecological farming system. One international group, the International Federation of Organic Agriculture Movements (IFOAM), has unsuccessfully tried to develop a universal standard (see box, "International Standards for Organic Agriculture").

In the EU, labeling of organic plant products is governed by EU Regulation 2092/91 (enacted in 1993) and of organically managed livestock by EU Regulation 1804/99 (enacted in 2000). Plant and animal products and processed organic foods imported into the EU may be labeled organic only if they conform to the provisions of the regulation. The regulation sets minimum rules for production, labeling, and marketing for the whole of Europe, but each country is responsible for interpreting and implementing the rules, as well as enforcement, monitoring, and inspection (Kilcher et al., 2004).

The EU recognizes a national authority from each member state as the body that can certify organic products as complying with EU law. These bodies, in turn, approve other certification entities within their member states that can also certify organic products. Most countries follow government-approved private certification bodies (table 4). The EU Regulation does not give specific guidance as to how member states should approve and supervise the certification bodies, resulting in a great deal of diversity among the states. Some countries and certifiers have additional public or private standards, particularly with regard to animal production and the use of inputs. Some also have standards for products not covered under the EU Regulation, such as fish and nonfood agricultural products (Rundgren, 2002).

International Standards for Organic Agriculture

According to the International Federation of Organic Agriculture Movements (IFOAM), based in Germany but active worldwide, organic agriculture "…..includes all agricultural systems that promote the environmentally, socially and economically sound production of food and fibres. These systems take local soil fertility as a key to successful production. By respecting the natural capacity of plants, animals and the landscape, it aims to optimise quality in all aspects of agriculture and the environment. Organic agriculture dramatically reduces external inputs by refraining from the use of chemo-synthetic fertilisers, pesticides, and pharmaceuticals. Instead it allows the powerful laws of nature to increase both agricultural yields and disease resistance. Organic agriculture adheres to globally accepted principles, which are implemented within local social-economic, geoclimatical and cultural settings" (IFOAM, 2004a).

Two international frameworks exist for certification bodies and standard-setting organizations for organic products: the IFOAM Basic Standards, which were developed in the 1980s, and the Global Codex Alimentarius standards, for which development began in 1999. IFOAM is an international federation of more than 600 affiliate organizations in more than 100 member countries, which focuses on general purposes, recommendations, basic standards, and derogations. One part of IFOAM's mission is international equivalency of organic quality claims so that certificates issued by certifiers in one part of the world are immediately accepted in other parts of the world. To facilitate international equivalency, IFOAM established an accreditation program in 1992, which is managed by the International Organic Accreditation Service (IOAS). IFOAM's basic standards acknowledge that national legislation will take into account local conditions that may well be more detailed than the IFOAM Basic Standards but would, as a minimum, meet IFOAM standards. However, the IFOAM standards are more rigorous in some instances than the relevant national standards, such as those laid down by EU or U.S. legislation. In fact, many IFOAM-accredited certifiers are beginning to demand stricter standards than, for example, the EU legislation requires, which industry experts suggest may lead to a tiered structure of the international certification industry (Organic Trade Services, 2004).

The Codex Alimentarius standards "provide an internationally agreed framework for organic food moving in international trade. Where a disagreement may occur between countries about the equivalence of organic food, the Codex guidelines can be used as a reference in trade disputes at the WTO level" (IFOAM, 2004b). For a few years, however, the Codex guidelines did not cover animal husbandry; livestock standards were approved in 2001.

Legally, if a producer or processor is certified by one of the EU-approved certification entities, certification should be valid in all EU member states. Some certification entities, however, insist on stricter standards than those of the EU legislation, which means that if the "brand" of the stricter certifier is strong in the marketplace, the producer or processor may find their customers insisting on that certifier's certification. Therefore, EU certification bodies have developed a tiered structure, whereby not all EU certificates are acceptable to all certification bodies (Organic Trade Services, 2004).

EU labeling of organic products is complex. In 2000, the EU introduced a voluntary logo for organic products that could be used throughout the EU by those meeting the regulation. At least 95 percent of the content of these products must originate in the EU. So far, few companies are using the logo (Dabbert et al., 2004; Kilcher et al., 2004). Some member states have public labels, while in other states, private certifiers have their own labels, some well known to the public (e.g., KRAV in Sweden, Skal in The Netherlands, or the Soil Association in the UK) (Rundgren, 2002). In most countries, private certifiers that have their own standards are legally obligated to certify producers to the EU Regulation if requested; in this case, the producer can use the EU label and member state label but not the private label of the certifier (Rundgren, 2002). The use of multiple labels can be confusing to the consumer, however. In some countries, only one logo is used and recognized. Other EU countries, however, use a range of labels, logos, and brands (Dabbert et al., 2004).

Table 4. EU organic certification

Country	Member state certification or government-approved private certification	Certification bodies
		Number
Austria	Private	8
Belgium	Private	2
Denmark	State/Plant Directorate, Food Directorate	
Finland	State	15 rural departments (for farms) and municipal food control (for processing)
France	Private	6
Germany	Private	22
Greece	Private	3
Ireland	Private	3
Italy	Private	13
Luxembourg	State and private	State inspects some small processors and 3 private
The Netherlands	Private	1 (Private certifier, Skal, has by law been given the status of government inspection authority and has a monopoly)
Portugal	Private	2
Spain	State and private	19 regional semi-governmental bodies and 1 private
Sweden	Private	2
United Kingdom	State (limited role) and Private(most certified by private)	10 private certifiers

Source: Rundgren, 2002;Organic Trade Services, 2004.

In 1990, USDAwas mandated through the Organic Foods Production Act (OFPA) to establish national standards for U.S. organic products. Up through the 1990s, and before any national standards were implemented, a patchwork of State laws existed regarding organic agriculture, with 17 States requiring certification for products labeled organic by the late 1990s. Thirteen other States had some type of regulation (although some had very minimal standards) but did not require third-party certification or inspection. Although certifiers each had their own standards and program management guidelines, Fetter and Caswell (2002) found in an analysis of the standards only minor differences among the certifiers, except in the area of livestock production. One major concern for the industry before the development of standards, however, was consumer confusion about what organic meant based on these standards, which was one of the primary motivations for developing national standards.

The three goals of OFPAas outlined in the legislation are to establish standards for marketing organically produced products, assure consumers that organic products meet a consistent standard, and facilitate interstate commerce (U.S. Congress, 1990). The legislation targeted environmental aspects by requiring that an organic production plan pay attention to soil fertility and regulate manure application to prevent water contamination and included environmental criteria to evaluate materials used in organic production (Merrigan, 2003; U.S. Congress, 1990). In practice, the official organic rule that was adopted also incorporated consumer desires to avoid GM ingredients and irradiation (Merrigan, 2003).

More than a decade after OFPA, the U.S. implemented the USDANational Organic Standards (NOS) on October 21, 2002. The Federal system still relies on multiple certifiers to certify producers and handlers of organic products, but these certifiers are now accredited by USDA. Although still too early to determine the overall impacts of the organic standard on the industry, the number of certifying agents has increased since implementation. The initial list of accredited organizations announced by USDAin April 2002 included 38 domestic and 4 non-U.S. organizations (Haumann, 2004). By March 2005, USDAlisted 56 domestic and 41 foreign accredited agents and 64 organizations as applicants. The growth in foreign agents is particularly notable. There is also evidence that U.S. consumers are becoming more aware of organic foods as more labeled products appear in stores (Haumann, 2004).

Conversion and Support Payments for EU Organic Farmers

The EU has proactively supported organic agriculture through a variety of schemes, including paying subsidies, or "green payments," to farmers for converting to organic farming and for continuing organic farming. No countries outside of the EU provide subsidies for existing or transitioning organic production. The economic rationale for green payments, which subsidize organic production, is that organic production provides benefits that accrue to society and that individual farmers do not consider these social benefits when making production decisions. In such cases, offering subsidies alters production practices by more closely aligning each farmer's private costs and benefits with social costs and benefits. In practice, green payments for organic production target new and existing organic farmers, partly to compensate new or "transitioning" farmers for the decline in yields when moving from conventional to organic production.[9]

The major policies for organic agriculture in the EU fall under the agri-environment program, the 1992 Common Agricultural Policy (CAP) reform (EC Regulation 2078/92), which provided the policy framework for EU member states to support organic farming. Many of the subsidies currently granted by European countries were implemented under this reform, and most countries implemented their schemes in 1994. Most of the countries have uniform national policies, although some have regional variations in the rates of payment and requirements. In fact, most countries have changed the original agri-environmental schemes for various reasons (Lampkin et al., 1999). More recently, under Agenda 2000, these measures were included in the rural development program (Rural Development Regulation No. 1257/99), a CAP reform carried out from 1999 to 2001.

As Dabbert and colleagues (2004) note, the response from regions and countries to devising the agri-environmental program has been diverse, and applying the program to organic farming even more so. Most of the schemes allow for newly converting and existing organic farmers to qualify for aid, with only France (with some regional exceptions) and the UK not supporting existing producers. Most countries allow staged conversions (with limits in some countries), where parts of the farm are converted over a number of years, and partial farm conversions. All schemes require that organic management of crops on the farm be maintained for 5 years, or payments must be refunded; more than half the countries do not require livestock to be managed organically. Several countries operate maximum and minimum size limits (by land area or business size), and half impose stocking rates. Afew countries restrict payments to specific crops.

The basic approach of the agri-environmental program is that farmers enter into a voluntary contract with a government agency, usually a 5-year contract, which commits the farmer to specific farming practices perceived as beneficial to the environment in return for payment (Dabbert et al., 2004). In 2001, almost €500 million were spent on organic lands under the two measures (2078/92 and 1257/99), with organic farms averaging payments of €183-€186 per hectare compared with €89 per hectare for conventional farms (table 5). Of this amount, about €270 million came from the Council Regulation 1257/99 measures, accounting for almost 16 percent of the total expenditures of €1.7 billion for 1257/99 (Häring et al., 2004).

Table 5. EU agri-environmental support and organic farming, 2001

Country	Organic land supported under agri-environmental programs 2078/92[1]	1257/99[2]	Share of organic land in policy support programs	Public expenditures for support of organic land under EC 2078/92and 1257/99	Average support premium for organic land 2078/92	1257/99
	Hectares		*Percent*	*Thousand*	*Euros/hectare*	
Austria	36,193	210,833	89	67,905	211	286
Belgium	13,032	3,616	74	3,416	187	269
Denmark	79,731	78,347	94	16,377	137	199
Finland	23,948	113,631	93	3,402	141	117
France	54,727	82,508	33	23,951	196	188
Germany	278,884	254,715	84	84,477	154	163
Greece	4,928	10,614	50	17,505	401	445
Ireland	13,691	NA	46	1,848	135	NA
Italy	351,113	101,134	37	158,898	361	318
Luxembourg	736	1,224	98	328	158	173
The Netherlands	8,140	14,593	63	4,446	266	156
Portugal	26,970	90	38	3,779	137	111
Spain	142,591	112,554	53	14,544	69	195
Sweden	81,067	349,562	113	69,018	153	162
United Kingdom	285,633	122,330	60	27,591	42	45
European Union	1,401,384	1,455,751	62	497,485	186	183

NA = Not available.

[1]Organic support falls under EC Regulation 2078/92, the agri-environment program, of the 1992 Common Agricultural Policy reform.See text for more detail.

[2]After 1999, organic farming support was part of Rural Development Regulation 1257/97, under Agenda 2000.

Source: Häring et al., 2004;Lampkin, 2005b.

The participation of organic farmers (measured by the amount of organic hectares supported by agri-enviromental programs) in the various EU countries demonstrates the differences among the countries in their support of organic lands. Dabbert and colleagues (2004, p. 45) note that, in most countries, "the organic land supported by the agri-environmental policies is a substantial proportion of the total certified organic area, which already points to the importance of these programmes for the development of organic farming in Europe." The share of organic land supported by agri-environmental policies ranges from 33 and 37 percent in France and Italy to almost 93 and 94 percent in Finland and Denmark (table 5). Sweden's 113 percent has more policy-supported organic land than certified area, reflecting Sweden's policy of supporting uncertified organically managed lands (see note 1, table 2 for more details).

TARGETS FOR EU LAND UNDER ORGANIC MANAGEMENT

Targets for land under organic production are used by EU member countries, and the organic farming action plans developed by several European countries provide targets for adoption of organic hectares by a certain year (e.g., 20 percent by 2010). The economic rationale for setting targets, in general, is that there is a threshold level above which a benefit accrues to society. Targets are used for farmland preservation in the U.S., for example, and for pollution emissions. In the case of targets for farmland preservation, the selected acreage target depends on the amount of land thought necessary to sustain a viable agricultural sector. If the target is not reached, the expectation is that farm-supporting industries (e.g., equipment and implement dealers, veterinary services, and credit institutions) will become economically stressed in the area because of the lack of farm operations. In turn, the local economy may be hurt by the loss of businesses, and the speed of farmland conversion to other uses may increase.

Organic farmland targets in the EU are used differently than farmland preservation targets in the U.S. The EU governments use targets to convey their level of commitment to growth in the organic sector. Some countries have selected relatively attainable targets, while others have chosen more ambitious ones.[10] For example, in 1995, Denmark announced a target of 7 percent of farmland certified as organic by 2000 and nearly reached this goal with 6 percent. Denmark's goal of having 12 percent of farmland certified as organic by 2003, however, fell short (table 6). In response to the 2000 BSE crisis, Germany set a target of certifying 20 percent of farmland as organic by 2010, a number that may be hard to reach with only 4 percent of land in organic production in 2003. Other countries have set varying targets: The Netherlands, 10 percent of farmland by 2010 and 5 percent of organic retail sales by 2005; Wales, 10 percent of farmland by 2005; France, 3 percent of farmland and 25,000 farms under organic management by 2005; Sweden, 20 percent of farmland in 2005; and Belgium, 10 percent of farmland by 2010.

State of Publicly Funded Research and Other Programs on Organic Agriculture

As markets continue to grow, public funding of organic-related research is increasing in both regions, although European governments are funding more organic research. EU and U.S. research agendas reflect the different priorities of the governments, with European funding focusing on innovation in production technique, food processing, food marketing, and food retailing. Niggli (2005) estimates that European state funding for organic farming research is €70-€80 million annually. However, countries vary widely in the amount of, and their approach to, funding for organic research (see Slabe, 2004, for a detailed accounting of programs and funding levels). Germany, The Netherlands, Switzerland, and Denmark are at the forefront of the research, accounting for 60 percent of all national spending on organic research in Europe (Niggli, 2005).

In the U.S., university-based technical assistance, Federal cost-share funds, and other State and Federal support for organic farmers and handlers are beginning to emerge. Nine USDAagencies have started or expanded programs on organic agriculture during the last several years (see Dimitri and Greene, 2002, appendix A, for more information). The Farm Security and Rural Investment Act of 2002 (commonly referred to as the 2002 Farm Bill) contained several first-time research and technical assistance provisions to directly assist organic crop and livestock producers with production and marketing. These provisions included a cost-share program to reimburse producers being certified and new organic research, education, and extension activities with a grants program. Also, several other provisions in the Act indirectly affect organic crop and livestock producers. For example, processes used to produce agricultural commodities, such as organic management, are now included in the definition of products that qualify for value-added market development grants. Several conservation assistance programs may also interest organic farmers. One—Agricultural Management Assistance—now specifically mentions organic farming among the practices that qualify for assistance to mitigate risk through market diversification and resource conservation practices.

In fiscal year 2005, about $7 million was made available exclusively for organic programs (e.g., the National Organic Program, certification cost-share program, and Integrated Organic Program). Of this amount, $4.7 million was for a first-time research grant program on organic agriculture. This amount, however, does not include money for other programs that benefit organic producers (such as USDA's Sustainable Agriculture Research and Education program and Value-

Added Producer Grants program), specific earmarks for organic research (e.g., those made to the Agricultural Research Service, USDA), and technical assistance and research by Federal, State, and local agencies. Nevertheless, these amounts demonstrate that U.S. support for organic agriculture is still small compared with that of the EU.

Table 6. EU agricultural land and organic targets

Country	Total agricultural land in country, 2002	Share of farmland that is organic, 2003	Policy targets for organic farmland
	1,000 hectares	*Percent*	*Percent and year*
Austria	3,387	9.7	NA
Belgium	1,393	1.7	10 by 2010 (in Flanders)
Denmark	2,690	6.1	7 by 2000;12 by 2003
Finland	2,216	7.2	10 by 2006;15 by 2010
France	29,622	1.9	3 by 2005
Germany	16,971	4.3	20 by 2010
Greece	3,917	6.2	NA
Ireland	4,372	.7	20 by 2010
Italy	15,341	6.9	NA
Luxembourg	127	2.4	NA
The Netherlands	2,933	2.2	10 by 2010
Portugal	3,813	3.2	NA
Spain	25,554	2.8	NA
Sweden[1]	3,039	7.4	20 by 2005
United Kingdom	15,722	4.4	Wales:10 by 2005; England:No targets for production

NA = Not available;data as of May 31, 2005.

Note: All data include organic and in-conversion lands.

[1]In Sweden, the percentage of organic farmland (7.4 percent) is for certified organic hectares only.However, Sweden also provides green payments to farmers of noncertified organic lands, and the policy target most likely refers to both certified and noncertified land.When the latter is included, 13.4 percent of the land is in either certified organic or noncertified organic lands.

Sources: Lampkin, 2005a;Häring et al., 2004;FiBL, 2005.

Recent Policies Related to Organic Agriculture

The most recent policy related to organic EU agriculture is the European Action Plan for Organic Food and Farming, adopted by the European Commission in June 2004. The plan seeks to facilitate ongoing development of organic farming and sets out a list of 21 policy actions to be implemented. The actions are focused on three main areas:

- Information-led development of the organic food market, including increasing consumer awareness, stimulating the use of the EU logo, and improving the availability of production, supply, and demand statistics.
- Making public support for organic farming more effective by encouraging member states to make a more coherent and greater use of the different rural development measures.
- Improving and reinforcing the EU's organic farming standards and import and inspection requirements.

Many have noted that, in the plan, the European Commission has a clearly stated rationale for promoting organic agriculture in the context of European agricultural policy. Some, however, have criticized its lack of targets for developing organic farming, the absence of new financial resources for implementation, and the fact that it does not consider the interaction of organic farming with the main components of the CAPand the new Rural Development Programmes (Stolze, 2005).

The impact of dramatic changes in the CAP—CAPReform 2003—on organic agriculture in Europe is still unknown. The reform shifts agricultural policy to a market-driven policy. One aspect of the reform is shifting to a single-farm payment independent of the farm production level. The farm payment will require cross compliance with a wide range of standards, including environmental, food safety, animal welfare, and occupational health and safety standards. The overall changes, however, are expected to favor an expansion of organic farming (Häring et al., 2004). Single-farm payments will begin in 2005-07, and member states will have significant discretion in implementing single-farm payments.

The most recent U.S. policy effort is the development of an Organic Caucus in the House of Representatives in 2002. The Caucus is a bipartisan group of 38 representatives (as of March 2005) with a mission to promote sound policies that advance organic production and marketing. The Caucus is likely to have some influence on the inclusion of organic agriculture policies in the next national farm bill. Another significant factor likely to shape U.S. organic policy is the 2005 ruling in the case, Harvey vs. the Secretary of Agriculture, which found that USDA's implementation of the organic standard is inconsistent with the Organic Foods Production Act of 1990. The ruling refers to USDA's allowing the use of synthetic substances in manufactured food products labeled as organic and to the practice of allowing herds transitioning to organic to consume 80 percent organic feed during the first 9 months of transition (see Klonsky and Greene, 2005, for a detailed discussion of this issue).

Discussions between the EU and U.S. have been ongoing with regard to equivalency between the organic standards of the two regions. The EU and U.S. have not yet reached an agreement about the equivalency of organic standards. However, the U.S. has recognition agreements with two European nations, the UK and Denmark, in which accredited agents of these two governments can certify that products were produced or handled in accordance with USDA regulations (USDA, AMS, 2005).

Market-Led or Government-Facilitated: Does It Matter?

While the EU and U.S. governments undoubtedly vary in their level and types of direct policy support of organic agriculture, we can say unequivocally that organic agriculture is dynamic in both regions. Some producers clearly choose organic farming methods for philosophical reasons, but others adopt organic methods because government policies encourage them to do so, particularly in Europe. Others are entering the organic sector in reaction to the market's price premiums and ostensibly higher profits. Retail sales in both regions also continue to expand quickly, although growth rates have recently slowed in Europe. Some consumers have begun purchasing organic food in response to food scares, others see organic food as a way of improving their health, while still others purchase for altruistic reasons, such as improving the environment. As a result of supply and demand growth, nearly all consumers in the developed world are aware of organic food, and many have better access to organic food in either specialized shops or mass-market stores.[11]

The one policy common to both the EU and U.S. is government-defined organic standards, which provide information to buyers about the unobservable characteristic, "organic." At this point, the commonalities in major policies between the two regions disappear. The EU has a wide variety of policies designed to increase the amount of land farmed organically, in keeping with their notion that organic farming provides benefits to society. The programs include green payments (subsidies), demandside policies, and land targets for organic production. U.S. funding for organic production and marketing has been limited, and although new programs have recently emerged, government efforts have resulted from intensive lobbying by the organic industry, unlike in the EU, where the governments actively support organic agriculture.

Viewing organic farming as providing public goods offers an economic rationale for government intervention in the market, which many European governments do. Viewing organic food as a differentiated product, created by using an ecological production system, suggests that a government should only regulate if there is concern about consumers being able to identify the product as "organic" or to reduce transaction costs of doing business, which is the approach taken by the U.S. Government.

Comparing market growth—both on the supplyside and demandside—in the context of these different policy choices is revealing. The retail market for organic products in Europe is larger, but the U.S. market is growing at a faster rate. In Europe, evidence suggests that, historically, supply has rapidly grown and, at different times, has needed to allow consumers to catch up. In the U.S., the opposite appears to be the case: Burgeoning consumer demand has pulled suppliers into organic production, so the U.S. market has had to cope with an industry that has frequently been supply constrained. Clearly, Europe has more (in both absolute value and as a share of the total) organic farms and farmland than does the U.S., which is likely the result of different histories and proactive supply-based government policies. Thus, the state of the organic sector in the two regions suggests that divergent policy directions do not matter in terms of meeting consumer demand, but they do matter in terms of satisfying other rationales for supporting organic agriculture.

ACKNOWLEDGMENTS

The authors thank Stephan Dabbert (University of Hohenheim), Nicolas Lampkin (University of Wales), John Love (USDA, World Agricultural Outlook Board), Kelly Strzelecki (USDA, Foreign Agricultural Service), Keith Jones (USDA, Agricultural Marketing Service), and Joseph Cooper, Catherine Greene, Joy Harwood, David Kelch, Barry Krissoff, and Robbin Shoemaker (USDA, Economic Research Service) for their insightful reviews of the manuscript. Appreciation is also extended to Linda Hatcher (USDA, Economic Research Service) for editorial and production support.

BIBLIOGRAPHY

Barry, Michelle. 2004. "The New World Order: Organic Consumer Lifestyle Segmentation," *[N]sight* VI(2). Bellview, WA: The Hartman Group, Inc.

Bertramsen, Sherrie, and Thomas Dobbs. 2002. "An Update on Prices of Organic Crops in Comparison to Conventional," *Economics Commentator*. South Dakota State University.

Conford, Phillip. 2001. *Origins of the Organic Movement*. Floris Books. September.

Dabbert, Stephan, Ann Maria Häring, and Raffaele Zanoli. 2004. *Organic Farming: Policies and Prospects*. Zed Books.

Dimitri, Carolyn, and Catherine Greene. 2002. *Recent Growth Patterns in the U.S. Organic Foods Market*. Agriculture Information Bulletin No. 777. U.S. Department of Agriculture, Economic Research Service, September. Available at http://www.ers.usda.gov/publications/aib777/

Dimitri, Carolyn, and Nessa J. Richman. 2000. *Organic Food Markets in Transition*. Public Policy Report No. 14. Henry A. Wallace Institute for Alternative Agriculture, April.

Fetter, Robert, and Julie Caswell. 2002. "Variation in Organic Standards Prior to the National Organic Program," *American Journal of Alternative Agriculture* 17(2):55-74.

Forschungsinstitut für biologischen Landbau (FiBL). 2005. *Organic Farming in Europe - Country Reports*. Available online at http://www.organic-europe.net/country_reports/default.asp

Foster, Carolyn, and Nicolas Lampkin. 1999. *European Organic Production Statistics 1993-1996. Organic Farming in Europe: Economics and Policy*, Volume 3. Universitat Hohenheim.

Glaser, Lewrene, Barry Krissoff, Angela Lengyel, and Kristen Sheehan. 1998. "Demand for Frozen Vegetables: AComparison of Organic and Conventional Products," *Vegetables and Specialties Situation and Outlook*. VGS-276. U.S. Department of Agriculture, Economic Research Service, November.

Glaser, Lewrene, and Gary Thompson. 2000. "Demand for Organic and Conventional Beverage Milk." Paper prepared for the Western Agricultural Economics Association meeting.

Greene, Catherine, and Amy Kremen. 2003. *U.S. Organic Farming in 2000-2001: Adoption of Certified Systems*. Agriculture Information Bulletin No. 780. U.S. Department of Agriculture, Economic Research Service, February. Available at http://www.ers.usda.gov/publications/aib780/

Grolink. 2002. "Overview of Implementation of the EU Organic Regulatory System," *The Organic Standard* 9, January.

Hallam, David. 2003. "The Organic Market in OECD Countries: Past Growth, Current Status, and Future Potential," *Organic Agriculture: Sustainability, Markets and Policies.* CABI Publishing and Organisation for Economic Co-operation and Development.

Hamm, Ulrich, and Gronefeld Friederike. 2004. *The European Market for Organic Food: Revised and Updated Analysis*, Volume 5 of the series "Organic Marketing Initiatives and Rural Development." University of Wales Aberystwyth, School of Management and Business.

Häring, Anna Maria, Stephan Dabbert, Joachim Aurbacher, Barbara Bichler, Christian Eichert, Danilo Gambelli, Nicolas Lampkin, Frank Offermann, Santiago Olmos, Jon Tuson, and Raffaele Zanoli. 2004. *Organic Farming and Measures of European Agricultural Policy. Organic Farming in Europe: Economics and Policy Volume 11.* Universitat Hohenheim.

The Hartman Group, Inc. 2002. *Hartman Organic Research Review.* Bellevue, WA.

Haumann, Barbara. 2004 "United States," *The World of Organic Agriculture 2004: Statistics and Emerging Trends*, Eds. Helga Willer and Minou Yussefi. Bonn, Germany: International Federation of Organic Agriculture Movements.

International Federation of Organic Agriculture Movements (IFOAM). 2004a. *Basic Standards.* Accessed December 2004 at http://www.ifoam.org/standard/basics.html#6

International Federation of Organic Agriculture Movements (IFOAM). 2004b. Accessed December 2004 at http://www.ifoam.org/fao/cod_alim.html

Kilcher, Lukas, Beate Huber, and Otto Schmid. 2004. "Standards and Regulations" *The World of Organic Agriculture 2004: Statistics and Emerging Trends*, Eds. Helga Willer and Minou Yussefi. Bonn, Germany: International Federation of Organic Agriculture Movements.

Klonsky, Karen, and Catherine Greene. 2005. "Widespread Adoption of Organic Agriculture in the U.S.: Are Market-Driven Policies Enough?" Paper prepared for American Agricultural Economics Association Annual Meeting.

Kortbech-Olesen, Rudy. 2003. "Overview on World Trade in Organic Food Products, the US Market and Recent Trends," Speech given at Biofach Congress, February 14, Nuremberg, Germany.

Lampkin, Nicolas. 2005a. *Certified and Policy-Supported Organic and In-Conversion Land Area in Europe (ha).* Data table. Institute for Rural Studies, University of Wales. Accessed April 2005 at http://www.organic.aber.ac.uk/statistics/europe05.html

Lampkin, Nicolas. 2005b. *Number of Certified and Policy-Supported Organic and In-Conversion Farms in Europe*. Data table. Institute for Rural Studies, University of Wales. Accessed April 2005 at http://www.organic.aber.ac.uk/statistics/europe05.html

Lampkin, Nicolas. 2005c. *Organic Farming in Europe - Statistics per 31.12.2004*. Available at http://www.organic-europe.net/europe_eu/statistics.asp

Lampkin, Nicolas. 2003. "From Conversion Payments to Integrated Action Plans in the European Union," *Organic Agriculture: Sustainability, Markets and Policies*. CABI Publishing in conjunction with the Organisation for Economic Co-operation and Development.

Lampkin, Nicolas, Carolyn Foster, Susanne Padel, and Peter Midmore. 1999. *The Policy and Regulatory Environment for Organic Farming in Europe. Organic Farming in Europe: Economics and Policy Volume 1*. Universitat Hohenheim.

Liebhardt, Bill. 2001. "Get the Facts Straight: Organic Agriculture Yields are Good." *Information Bulletin* 10(Summer). Organic Farming Research Foundation.

Merrigan, Kathleen. 2003. "The Role of Government Standards and Market Facilitation," *Organic Agriculture: Sustainability, Markets and Policies*. CABI Publishing in conjunction with the Organisation for Economic Cooperation and Development.

Michelsen, J., U. Hamm, E. Wynen, and E. Roth. 1999. "The European Market for Organic Products: Growth and Development," *Organic Farming in Europe: Economics and Policy Volume 7*. Universitat Hohenheim.

Niggli, Urs. 2005. "Organic Farming Research—Definitely Out of the Niche," *Organic Farming in Europe 2005: Market, Production, Policy & Research*, Ed. Nicolas Lampkin. Available at http://www.fibl.org

Nutrition Business Journal. 2003. *The NBJ/SPINS Organic Foods Report 2003*. Penton Media, Inc.

Nutrition Business Journal. 2004. *The NBJ/SPINS Organic Foods Report 2004*. Penton Media, Inc.

Oberholtzer, Lydia, Carolyn Dimitri, and Catherine Greene. 2005. *Price Premiums Hold on as U.S. Organic Produce Market Expands*. VGS-308-01. U.S. Department of Agriculture, Economic Research Service, May. Available at *http://www.ers.usda.gov/publications/vgs/may05/VGS30801/VGS30801.pdf*

Offermann, Frank, and Hiltrud Nieberg. 2000. "Economic Performance of Organic Farms in Europe," *Organic Farming in Europe: Economics and Policy Volume 5*. University of Hohenheim.

Organic Business News. 2004. "High Demand for Dairy Products Creates Shortage for Organic Suppliers," 16(12), December.

Organic Monitor. 2003. *The Global Market for Organic Food and Drink.*

Research Publication #7001-40. July. Available at http://www.organicmonitor.com

Organic Monitor. 2005. *Organic Milk Shortage Imminent.* March 11. Available at http://www.organicmonitor.com

Organic Trade Association. 2004. *Organic Trade Association's 2004 Manufacturer Survey*. Produced by Nutritional Business Journal. May.

Organic Trade Services. 2004. Available at http://www.organicts.com/index.html

Richter, Toralf. 2005. "The European Market Between Strong Growth and Consolidation-Current State and Prospects," *Organic Farming in Europe 2005: Market, Production, Policy & Research*, Ed. Nicolas Lampkin. Available at http://www.fibl.org

Robinson, Barbara. 2003. "Value Through Verification. USDA National Organic Program." U.S. Department of Agriculture, Agricultural Marketing Service. Available at *http://www.ams.usda.gov/nop/NOPPresentation/Introduction.html*

Rundgren, Gunnar. 2002. "Overview of the Implementation of the EU Organic Regulatory System," *The Organic Standard* 9, January, pp. 3-9.

Schneider, Monika, Toralf Richter, Christoph Spahn, and Katrin Portmann. 2005. *Overview of International Organic Market Development and Potential Export Markets for Organic Products of Ukraine.* FiBLProject Report. Research Institute of Organic Agriculture (FiBL).

Slabe, Anamarija. 2004. *Organic Food and Farming Research in Europe: Consolidated Report—How to Improve Trans-National Cooperation.* Prepared for the Organic Food and Farming Research in Europe: How to Improve Trans-National Cooperation conference in Brussels, Belgium, November 26. Available at http://europa.eu.int/comm/research/agriculture/pdf/off_research_in_europe_ consolidate_report.pdf

Sok, Emy, and Lewrene Glaser. 2001. "Tracking Wholesale Prices for Organic Fruit," *Agricultural Outlook* 285:7-8. U.S. Department of Agriculture, Economic Research Service, October.

Stolze, Matthias. 2005. *The Current Agri-Policy Context: The European Action Plan for Organic Farming and the Current CAPReform.* Presentation at BioFach Congress 2005, Nuremburg, Germany. Available at http://www.fibl.org

Streff, Nicholas and Thomas Dobbs. 2004. *"Organic" and "Conventional" Grain and Soybean Prices in the Northern Great Plains and Upper Midwest: 1995 through 2003.* Econ Pamphlet 2004-1. South Dakota State University Economics Department, June.

U.S. Congress. 1990. The Organic Foods Production Act, Title XXI of the Food, Agriculture, Conservation, and Trade Act of 1990 (Public Law 101-624).

U.S. Department of Agriculture, Agricultural Marketing Service (USDA, AMS), National Organic Program. 2002. "Organic Food Standards and Labels: The Facts." Available at http://www.ams.usda.gov/nop/Consumers/brochure.html

U.S. Department of Agriculture, Agricultural Marketing Service (USDA, AMS), National Organic Program. 2005. "Trade Issues." Available at http://www.ams.usda.gov/nop/NOP/Trade.html

U.S. Department of Agriculture, Foreign Agricultural Service (USDA, FAS). 1999. FAS GAIN Report GM9071.

U.S. Department of Agriculture, Foreign Agricultural Service (USDA, FAS). 2002. FAS GAIN Report GM2002.

U.S. Department of Agriculture, Foreign Agricultural Service (USDA, FAS). 2005. FAS GAIN Report E35991.

U.S. Department of Agriculture, Foreign Agricultural Service (USDA, FAS). 2005. *U.S. Market Profile for Organic Food Products*. February.

Walnut Acres. 2001. *The Walnut Acres Certified Organic Future*. Available at http://www.walnutacres.com

Walz, Erica. 2004. *Final Results of the Fourth National Organic Farmers' Survey: Sustaining Organic Farms in a Changing Organic Marketplace*. Organic Farming Research Foundation. Available at http://www.ofrf.org/publications/survey/index.html

Whole Foods. 2004. *Organic Foods Continue to Grow in Popularity According to Whole Foods Market Survey.* October. Available at http://www.wholefoods.com/company/pr_10-21-04.html

Willer, Helga, and Bernward Geier. 2005. "More than 26 Million Certified Organic Hectares Worldwide." Press release issued by the International Federation of Organic Agriculture Movements, FiBL, SOL, NurnbergMesse.

Willer, Helga, and Toralf Richter. 2004. "Europe," *The World of Organic Agriculture 2004: Statistics and Emerging Trends*, Eds. Helga Willer and Minou Yussefi. Bonn, Germany: International Federation of Organic Agriculture Movements.

Willer, Helga, and Yussefi, Minou, Eds. 2004. *The World of Organic Agriculture 2004: Statistics and Emerging Trends.* Bonn, Germany: International Federation of Organic Agriculture Movements.

REFERENCES

[1] Conversion from euros to dollars uses the exchange rate current in May 2005, €1.00 = $1.29.

[2] The EU-15 countries (the countries that made up the EU before entry of 10 new countries in May 2004) are the focus of this report because much of the historical data on organic agricul- ture in the EU is based on these coun- tries. There are recent efforts to collect more information on organic produc- tion and markets in the 10 new coun- tries (see Lampkin, 2005a-c). All references to the EU in this report refer to the EU-15. The 15 countries are Austria, Belgium, Denmark, Finland, France, Germany, Greece, Ireland, Italy, Luxembourg, The Netherlands, Portugal, Spain, Sweden, and the United Kingdom.

[3] The most recent statistics on U.S. organic farmland are for 2001.

[4] The organic sector is not immune to food scares; in 2002, organic wheat—used as food for poultry by German farmers—was contaminated by a weed killer, nitrofen, which is banned by Germany. The grain was contaminated in a warehouse that had previously stored herbicides. This incident dampened German consumer enthusiasm for organic food.

[5] See Hamm and Gronefeld (2004) for an in-depth examination of the European markets for the year 2001.

[6] We thank Stephan Dabbert for pointing this out to us.

[7] The two regions with the greatest amount of certified organic farmland are Australia (42 percent of total farmland) and Latin America (24 percent). Alarge share of this is pastureland used for raising livestock (Willer and Yussefi, 2004).

[8] We thank Nicolas Lampkin and Stephan Dabbert for pointing out this EU debate.

[9] U.S. farmers do not experience a similar post-transition yield reduction (see Liebhardt, 2001, for more infor- mation), although yields for major crops may be lower over time because of organic crop rotation. Organic EU farmers, on the other hand, often con- tinue to have yields below conven- tional producers after the 3-year conversion period. For example, organic grain yields are about 60-70 percent of conventional grain yields (Offermann and Nieberg, 2000). One possible explanation for this difference is that, for conventional production systems, land is farmed more inten- sively in Europe than in the U.S.

[10] The authors are grateful to Stephan Dabbert for clarifying the rationale for organic land targets.

[11] Access to organic food is not without challenges: Finding organic food in restaurants, for example, in both the EU and U.S. is difficult. Organic food is more readily available in foodstores in some U.S. States and in some EU countries than in others.

In: Organic Agriculture in the U.S.
Editor: Alison J. Wellson, pp. 129-190
ISBN: 978-1-60021-305-2

Chapter 5

RECENT GROWTH PATTERNS IN THE U.S. ORGANIC FOODS MARKET

Carolyn Dimitri and Catherine Greene

ABSTRACT

Organic farming is one of the fastest growing segments of U.S. agriculture. As consumer interest continues to gather momentum, many U.S. producers, manufacturers, distributors, and retailers are specializing in growing, processing, and marketing an ever-widening array of organic agricultural and food products. This report summarizes growth patterns in the U.S. organic sector in recent years, by market category, and describes various research, regulatory, and other ongoing programs on organic agriculture in the U.S. Department of Agriculture.

Keywords: organic agriculture, organic farming systems, organic marketing, organic marketing channels, certified organic acreage and livestock, price premiums, national organic rules, specialty agriculture, high-value crops, USDA research.

SUMMARY

Burgeoning consumer interest in organically grown foods has opened new market opportunities for producers and is leading to a transformation in the organic foods industry. Once a niche product sold in a limited number of retail outlets, organic foods are currently sold in a wide variety of venues including farmers markets, natural product supermarkets, conventional supermarkets, and club stores. Since the early 1990s, certified organic acreage has increased as producers strive to meet increasing demand for organic agricultural and food products in the United States. The dramatic growth of the industry spurred Federal policy to facilitate organic product marketing, and is leading to new government activities in research and education on organic farming systems.

This report summarizes growth patterns in the U.S. organic sector in recent years, by market category, and traces the marketing channels for major organic commodity groups. The report describes various research, regulatory, and other ongoing programs on organic agriculture in the U.S. Department of Agriculture.

- The U.S. organic food industry crossed a threshold in 2000: for the first time, more organic food was purchased in conventional supermarkets than in any other venue.
- Growth in retail sales has equaled 20 percent or more annually since 1990. Organic products are now available in nearly 20,000 natural foods stores, and are sold in 73 percent of all conventional grocery stores.
- According to the most recent USDA estimates, U.S. certified organic cropland doubled between 1992 and 1997, to 1.3 million acres.
- The new U.S. Department of Agriculture standards for organic food, slated to be fully implemented by October 2002, are expected to facilitate further growth in the organic foods industry.
- Fresh produce is the top-selling organic category, followed by nondairy beverages, breads and grains, packaged foods (frozen and dried prepared foods, baby food, soups, and desserts), and dairy products. During the 1990s, organic dairy was the most rapidly growing segment, with sales up over 500 percent between 1994 and 1999.
- Nine USDA agencies have expanded research, regulatory, and other programs on organic agriculture.
- The main regulatory program is the creation, implementation, and administration of the USDA organic standard. Other programs include crop insurance for organic farmers, information provision, and promotion of organic exports.
- USDA also funds projects for international market development and for natural resource conservation. Funding is also extended to projects assisting adoption of organic practices and exploration of new farming systems, methods, and educational opportunities.

- USDA research includes agronomic studies on soil management, biological control of pests and weeds, livestock issues, and post-harvest fruit treatment. Economic research focuses on tracking growth in the organic sector, demand for organic products, and organic farmers'risk management strategies.

INTRODUCTION

Burgeoning consumer interest in organically grown foods has opened new market opportunities for producers and is leading to a transformation in the organic foods industry. Once a niche product sold in a limited number of retail outlets, organic foods are currently sold in a wide variety of venues including farmers markets, natural product supermarkets, conventional supermarkets, and club stores. Many U.S. manufacturers and distributors are specializing in processing and marketing organic products, while some longtime manufacturers of conventional products have introduced organic items to their product lines. As a result, an ever-widening array of organic agricultural and food products is now available. Since the early 1990s, certified organic acreage has increased as producers strive to meet increasing demand for organic agricultural and food products in the United States. The dramatic growth of the industry spurred Federal policy to facilitate organic product marketing, and is leading to new government activities in research and education on organic farming systems.

This report summarizes growth patterns in the U.S. organic sector in recent years, by market category, and describes various research, regulatory, and other ongoing programs on organic agriculture in the U.S. Department of Agriculture.

MARKETPLACE CHARACTERISTICS OF U.S.ORGANIC SECTOR

The U.S. organic food industry crossed a threshold in 2000: for the first time, more organic food was purchased in conventional supermarkets than in any other venue. Industry estimates suggest that nearly half of the $7.8 billion spent on organic food in 2000 was purchased in conventional retail outlets. Organic products are now available in nearly 20,000 natural foods stores *(Natural Foods Merchandiser)*, and are sold in 73 percent of all conventional grocery stores (Food Marketing Institute).

Growing consumer demand for organic products has been manifested in the market in many ways. Acreage of certified organic farmland is increasing to meet growing consumer demand. According to the most recent USDA estimates, U.S. certified organic cropland doubled between 1992 and 1997, to 1.3 million acres. Preliminary estimates for 2001 suggest that certified organic acreage significantly increased between 1997 and 2001. From the consumer side, new products are being introduced rapidly. For example, over 800 new organic products were introduced in the first half of 2000. Desserts made up the majority of new products in 2000, while most new products introduced in 1999 were beverages (Myers and Rorie).

The new U.S. Department of Agriculture standards for organic food, slated to be fully implemented by October 2002, are expected to facilitate further growth in the organic foods industry. The USDA standard defines organic production as "A production system that is managed in accordance with the [Organic Foods Production] Act and regulations in this part to respond to site-specific conditions by integrating cultural, biological, and mechanical practices that foster cycling of resources, promote ecological balance, and conserve biodiversity."[1] The national organic standards address the methods, practices, and substances used in producing and handling crops, livestock, and processed agricultural products (see box). All agricultural products that are sold, labeled, or represented as organic must be in compliance with the regulations after October 2002.

Organic food is sold to consumers through three main venues in the United States—natural foods stores, conventional grocery stores, and direct-to-consumer markets—and a small amount is exported to foreign markets. USDA does not have nation- al statistics on organic retail sales. Industry sources have reported retail sales for organic food, but those data are fragmentary and, at times, inconsistent. A trade publication, the *Natural Foods Merchandiser* (NFM) reported estimates of total U.S. retail sales of organic foods for 1990 through 1996. NFM estimated total organic sales through all marketing outlets rose steadily from about $1 billion in 1990 to $3.3 billion in 1996, the last year that total sales were reported. Since 1999, Packaged Facts, a market research firm, has been reporting organic food sales. According to Packaged Facts, organic food sales in all venues totaled $6.5 billion in 1999 and $7.8 billion in 2000. This increase continues the streak of industry growth equal to 20 percent or more annually since 1990.

Natural Product and Conventional Supermarkets

From 1990 to 1996, natural products retailers sold two-thirds of organic foods. Sales of organic food in natural products stores increased 20 to 25 percent annually since the early 1990s. Natural product retailers comprise 1 percent of all food stores in the United States, and sold 48 per- cent of all organic food in 2000. Conventional food stores began outselling natural products supermarkets in several categories in 1999, including organic milk, half and half, cream, nondairy beverages, cold cereals, cookies and snack bars, and tofu. In 2000, conventional supermarkets comprised 99 percent of all food stores and sold 49 percent of all organic products.

Purveyors of natural products were the primary sales force for organic food since the beginning of the organic food movement over half a century ago. Until 2000, the largest retail out- let for organic food was natural foods stores followed by direct markets (such as farmers markets), according to NFM data (fig. 1). In 2000, 49 per- cent of all organic products was sold in conventional supermarkets, 48 per- cent was sold in health and natural products stores, and 3 percent through direct-to-consumer methods (Packaged Facts). In contrast, in 1991, 7 percent of all organic products were sold in conventional super- markets and 68 percent were sold in health and natural products stores (NFM).

Fresh produce remains the top-selling organic category (see fig. 2), followed by nondairy beverages, breads and grains, packaged foods (frozen and dried prepared foods, baby food, soups, and desserts), and dairy products. During the 1990s, organic dairy was the most rapidly growing segment, with sales up over 500 percent between 1994 and 1999. Sales of organic yogurt and kefir increased 56.4 percent between 1999 and 2000. Following closely, sales of nondairy beverages (for example, juice and soymilk) increased 53.1 percent and sales of fresh produce grew by 51.4 percent between 1999 and 2000, according to industry sources. Over- all, according to Packaged Facts, organic sales in natural product supermarkets and conventional stores increased by 20 percent between 1999 and 2000.

Organic farmers market their food directly to consumers much more frequently than conventional farmers do, and the last decade has seen a renaissance in the use of farmers markets across the country. Producers capture a much higher share of the consumer food dollar when they market their produce directly to consumers. Sever- al surveys of certified organic producers show similar findings on their heavy use of direct-to-consumer marketing. A 1997 survey of certified organic producers in the United States conducted by the Organic Farming Research Foundation (OFRF)—a California nonprofit group that sponsors research on organic farming—found that direct market use is extensive and varies by commodity sector, with fruits and vegetables the highest. Organic producers reported selling produce from about 23 percent of their vegetable acreage directly to consumers through on-farm sales (9 percent), farmers markets (8 percent), "community sup- ported agriculture" subscriptions (4 percent), and other types of direct-to- consumer markets (2 percent). Also, produce from about 20 percent of the organic fruit and vegetable acreage was marketed directly to grocery retailers and restaurants. A 1994 USDA survey of certified organic vegetable producers in the United States found that the use of direct-to- consumer markets varied with farm size, with 60 percent of the growers with under 10 acres (three-quarters of the respondents) using this channel compared with 12 percent with 10 acres or more (Fernandez-Cornejo et al.). Smaller growers tended to market directly to grocery retailers (11 percent versus 6 percent for larger growers) and through grower cooperatives (10 percent versus 3 percent for larger growers), while the larger growers marketed more heavily to vegetable packer/shippers, brokers, and food processors.

Organically grown food is widely available in farmers markets across the United States, and organic-only farmers markets have been organized in Oregon, Illinois, Missouri, and other States. The renaissance in farmers markets in the United States during the 1990s—fostered by State and local municipalities wanting to revitalize neighborhoods and preserve regional farmland and open space— has been a boon to organic farmers who use this marketing outlet much more heavily than conventional farmers do. States are also producing directories of farm stands and pick- your-own farms, including organic directories, and developing logos like "Jersey Fresh" to promote locally grown food.

Community-supported agriculture (CSA) is an innovative direct marketing arrangement that organic farmers have been pioneering in the United States for about a decade. Consumers subscribe to the harvest of a CSA farmer for the entire upcoming season, and pay for their produce in advance. Under a CSA arrangement, consumers share the production risks and variable harvests of the farmer—including especially abundant harvests—and sometimes participate in festivals and other social activities at the farm. Over 800 CSAs are currently listed in the U.S. database maintained by USDA and the Robyn Van En Center at Wilson College. Most of the CSA farms use organic production systems.

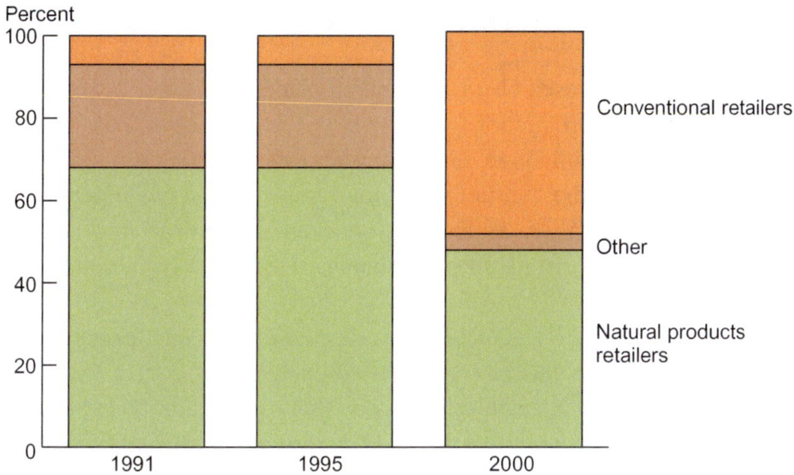

Notes: Other is direct sales and exports.
Source: Natural Foods Merchandiser, Packaged Facts.

Figure 1. Share of organic sales by venue

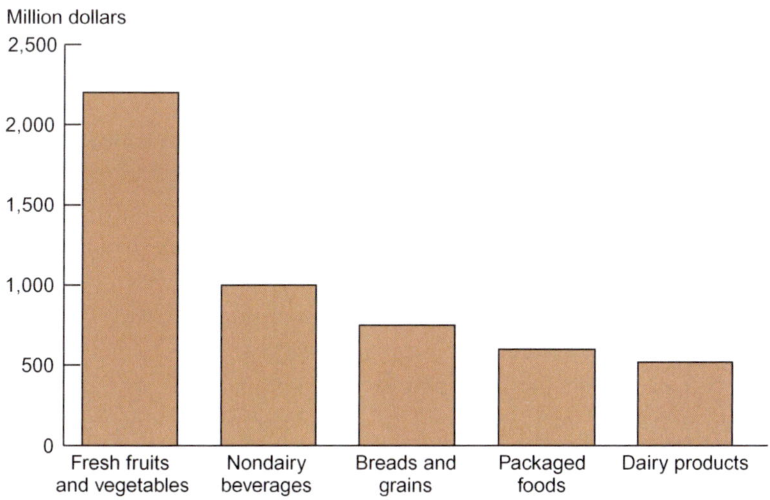

Source: Nutrition Business Journal.

Figure 2. Sales of top five organic foods, 2000

PRODUCTION CHARACTERISTICS OF U.S.ORGANIC SECTOR

A growing body of research in the United States has been devoted to the economics of organic production systems—its yields, input costs, income, profitability, and other economic characteristics. A 1990 review of the U.S. literature concluded that the "variation within organic and conventional farming systems is likely as large as the differences between the two systems" and found mixed results in the comparisons for most characteristics (Knoblauch, Brown, and Braster, 1990).

Several more recent U.S. studies have indicated that organic price premiums are key in giving organic farming systems comparable or higher whole- farm profits than conventional chemical-intensive systems, particularly for crops like processed tomatoes and cotton (Klonsky and Livingston, 1994; Batte, Forster, and Hitzhusen, 1993; Assadian, Esparza, and Ponce, 1999). Other studies have found that organic systems may be more profitable than conventional systems, even without price premiums. For example, some Midwestern organic grain and soybean production was found to be more profitable than conventional systems, even without price premiums, due to higher yields in drier areas or periods, lower input costs, or crop mix (Welsh, 1999). Also, a recent study comparing organic and conventional apple production in California's Central Coast showed higher yields as well as higher returns under the organic systems (Swezey et al., 1994). And another recent study compared organic, conventional, and integrated apple production systems in Washington State over a 6-year period, and found that the organic system was more profitable, had similar yields, better tasting fruit, and was more environmentally sustainable and energy efficient than the other systems (Reganold et al.). We are not aware of recently published research that finds farming with organic methods is less profitable than farming with conventional methods. Of course, net returns to various production systems may vary with biophysical and economic factors (such as soil type, climate, and proximity to markets), and a system that is optimal in one location may not be optimal in another. Also, factors not captured in standard profit calculations, such as convenience, longer-term planning horizons, and environmental ethics, can motivate rational adoption of a particular practice or farming system. Further research is needed to improve our understanding of the factors influencing net returns to organic farming systems.

The promising results from the limited number of economic studies to date have led to an increase in research on organic farming systems. USDA, universities, and other U.S. institutions are increasingly examining the long-term economics of organic farming systems through replicated field trial research and a multidisciplinary systems approach. Most of these projects are less than a decade old, and promise to answer basic research questions about yields and profitability as well as to address farmer-defined management and production obstacles to the more wide- spread adoption of organic production systems.

Examples of U.S. Long-Term Farming Systems Trials

Rodale Institute's Farming Systems Trial™, in Kutztown, Pennsylvania, was begun in 1981 and is one of the longest running experiments designed specifically to study organic cropping systems. The project focuses on corn and soybean production and studies the transition process that occurs when converting from conventional to organic farming.

The Sustainable Agriculture Farming Systems Project, at the University of California, Davis, has been examining processing tomato, safflower, bean, and corn production under conventional, low-input, and organic systems since 1988 on a 28-acre site in the Sacramento Valley.

Elwell Agroecology Farm, in conjunction with the University of Minnesota's Lamberton Experiment Station, began adding long-term, organic cropping systems trials in 1989;rotations include corn, soybeans, alfalfa, and oats.

The Living Field Laboratory, a project of the C.S.Mott Chair of Sustainable Agriculture, Michigan Agricultural Experiment Station, and others, was established in 1993 and is examining corn, soybean, and wheat production under organic, conventional, and other management systems.

USDA's Agricultural Research Service began investigating organic farming systems through long-term farming systems trials in 1993. The Farming Systems Project, at USDA's Beltsville (Maryland) Agricultural Research Center, is a long-term comparison of seven cropping systems with various rotations of corn, soybean, wheat, and alfalfa that are typical in the mid- Atlantic region.

The West Virginia University (WVU) converted its entire 60-acre Horticulture Farm to organic production in the fall of 1999. The WVU Organic Research Farming Project is studying market garden/vegetable production systems, as well as field crop/livestock systems, in replicated plots, and is evaluating changes in various aspects of the fauna, flora, and soil as organic practices are followed.

Farmers in 49 States used organic production methods and third-party organic certification services on 1.35 million acres of farmland in 1997, according to an Economic Research Service (ERS) study. Crops were grown on about two-thirds of the certified U.S. organic farmland, and the rest was pasture. The ERS study analyzed data from 40 State and private certifiers (see box, p. 8, on organic standards and certification). Uncertified acreage was excluded, even though it may represent a large segment of organic production, because of the difficulty in determining the production criteria used by uncertified growers. ERS reports statistics on certified organic U.S. acreage in the Organic Farming and Marketing Briefing Room *(www.ers.usda.gov/ briefing/organic).*

Organic farming has made deeper inroads in the fruit, vegetable, and other high-value specialty crop industries than in the major grain and oilseed industries. While less than two-tenths of 1 percent of the U.S. corn, soybean, and wheat crops were grown organically in 1997, over 1 percent of the dry peas and tomato crops and about 2 percent of the apple, grape, lettuce and carrot crops were organic. And nearly a third of the U.S. herb and "mixed vegetable" crops were grown organically in 1997. (A "mixed vegetable" crop is a mixture of numerous horticultural crops (mostly vegetables) grown on a small farm or parcel.)

The markets for organic vegetables, fruits, and herbs have been developing for decades in the United States, and these crops are grown organically in more States than any other type of commodity. State and private certifying groups certified over 180,000 acres of these crops in 44 States in 1997, more than double the amount certified in 1994, with the biggest gains for cultivated and wild-harvested herbs such as St. John's Wort.

About 2 percent of the major fruit and vegetable crops—apples, carrots, lettuce, and grapes—were grown organically, and a third of the organic vegetable acreage was devoted to producing "mixed vegetables" in 1997. Mixed vegetable farms, as defined in the census of agriculture, are small farms—less than 50 acres—that pro- duce a large number of vegetables. Large farms produce processing tomatoes, organic wine grapes, and other high-value crops on a commercial scale, while numerous small farms still specialize in mixed vegetable production for direct marketing to consumers and restaurants. The top producer of organic fruits and vegetables was California, followed by Arizona, Florida, Texas, and Washington.

About a third of the total certified organic vegetable acreage in 1997 was for mixed vegetables. In 1997, U.S. farmers certified nearly 3,000 acres of organic mixed vegetables on farms or parcels that were 5 acres or less, and over 14,000 acres on farms and parcels over 5 acres. New York organic producers had over 1,400 acres in the 5-acres-or-less category. Mixed vegetable producers often target farmers markets, community-sup- ported agriculture subscriptions, restaurants, and other direct marketing outlets.

Organic farmers are also growing major grains and oilseeds on a small portion of the planted area in the United States. Wheat was produced under certified organic farming systems on over 125,000 acres in 1997, corn was grown on over 42,000 acres, and soybeans were produced on about 82,000 acres. Other field crops produced organically in 1997 include barley, oats, sorghum, rice, spelt, mil- let, buckwheat, rye, dry peas, lentils, dry beans, flax, and sunflowers. Organic acreage of these crops, especially soybeans, has undoubtedly increased since 1997.

Thirty-nine States had certified organic hay and silage production, with most acreage in Idaho, Wisconsin, and New York. Acreage of these crops expanded 51 percent between 1995 and 1997 as the number of certified organic milk cows more than doubled during that period.

Organic meat and poultry markets have lagged behind those for crops partly because meat and poultry could not be labeled as organic until February 1999, when a provisional label was approved by USDA. Food crops and non-meat animal foods (eggs and dairy products) are regulated by the U.S. Food and Drug Administration, which allowed food packages to carry an organic label throughout the 1990s.[2] While the number of certified organic beef cows, hogs, sheep, and lambs declined during the study period (1992-97), the number of dairy cows and layer hens increased sharply. The market for organic meat products is beginning to grow now that organic labeling is permitted, and the growing market for organic milk and eggs has been pushing up the use of certified organic pasture and the demand for certified organic grains and oilseeds.

Farmers and ranchers raised a small number of certified organic cows, hogs, and sheep in 23 States in 1997. Dairy cows were raised organically in 13 States in 1997, and New York, Wisconsin, and Minnesota were the top three producers. The number of certified organic milk cows in the United States nearly tripled between 1992 and 1994 and more than doubled between 1994 and 1997. California was the leader in organic poultry production, followed by New York and Virginia. Other organic animal specialties, including goats, fish, and colonies of bees, were certified in several States.

Consumption Characteristics of the U.S. Organic Sector

A number of academic and industry studies have been conducted to examine consumer behavior and identify their motivation for purchasing organic foods. Many of the industry studies use consumer surveys, which seek to identify how often consumers purchase organic food, their motivations for purchasing organic food, and demographic data on organic food purchasers. University studies have adopted different approaches to assess consumer buying behavior and to identify which characteristics (for example, income, food quality, educational level, concern for the environment, or family size) affect whether consumers will purchase organic food.

Several industry groups have surveyed consumers about their preferences and buying habits for organic food. The results of the different surveys are not always consistent. *The Nutrition Business Journal* reported that 11 percent of consumers purchased some organic food in 2000, and less than 2 percent are regular purchasers. Results of the Hartman Group's 2000 survey suggest that 3 percent of consumers regularly buy organic products. The Walnut Acres Survey (2001) found that 63 percent of respondents purchased organic food at least sometimes, and 57 percent of the purchasers had been doing so for at least 3 years. The Food Marketing Institute's survey (2001) found that 66 percent of surveyed shoppers bought organically grown foods.

In 2001, the Food Marketing Institute's survey indicated that 37 percent of shoppers said they purchased organically grown food to maintain their health; and 44 percent of these shoppers had purchased organic food in the past 6 months. Consumers surveyed by the Hartman Group (2000) gave the following reasons for purchasing organic food: health and nutrition (66 percent), taste (38 percent), environment (26 percent), and availability (16 percent).[3] The Fresh Trends (2001) survey revealed that 12 percent of the shoppers surveyed reported that whether a product is organic is a primary factor in their purchasing decision. Sixty-three percent of the respondents of the Walnut Acres Survey believed that organic food and beverages were better for them and were more healthful than their conventional counterparts. Fresh Trends (1996, 1998, 2000, 2002) found little difference between the purchasing habits of men and women.

Over the years, Fresh Trends found that, of the shoppers that had purchased organic produce in the previous 6 months, more purchased vegetables than fruit (according to the 1996 survey, 24 percent purchased fruit and 84 percent purchased vegetables; according to the 2000 survey, 35 percent purchased vegetables). Apples and tomatoes led the list of fruit and vegetables purchased by the shoppers surveyed by Fresh Trends. According to the Hartman Survey (2000), the top 10 organic products purchased were strawberries, lettuce, carrots, other fresh fruit, broccoli, apples, other fresh vegetables, grapes, bananas, and potatoes. The Hartman survey also suggested that fruits and vegetables were "gateway categories" (typically the first organic products purchased by consumers). The Walnut Acres Survey (2002) indicated that 68 percent of consumers revealed that price is the main reason they did not purchase organic food.

Summary of Methodology Used in Consumer Studies

Study	Year Published	Methodology
Walnut Acres Survey	2001	Telephone interviews with nationally representative sample of 1,000 adults. Sample collected March 1 - March 5, 2001.
Food Marketing Institute Survey	2001	Telephone interviews with nationally representative sample of 1,200 adults.
Hartman Group, the Organic Consumer Profile	2000	Mail questionnaire sent to 40,000 households, nationwide, selected from a panel of 550,000 households. The panel conforms to a cross-section of the population based on the 2000 census; 26,434 consumers responded.
Fresh Trends Survey	Biannual	Two separate surveys. The first survey was a telephone survey of 1,000 nationally representative households. Average phone conversation was 10 minutes. The second survey covered 5,000 consumer in-store interviews in June 2001 and August 2001. Interviews took place in one retailer (five stores each) in five different markets across the United States.
Estes and Smith	1996	Hedonic analysis of demand for organic produce. Data collected using focus groups and mall interviews.
Thompson and Kidwell	1998	Collected data on items purchased by examining shopping carts of consumers in Tucson, AZ, and collected demographic data.
Glaser and Thompson	1999, 2000	Econometric analysis (using scanner data) of demand for organic frozen vegetables and organic milk.
Reicks, Splett, and Fishman	1999	Experimental design for effectiveness of shelf labeling.
Govindasamy and Italia	1999	Estimated consumer willingness to pay a 10-percent premium for organic produce.
Thompson and Glaser	2001	Econometric analysis (using scanner data) of demand for organic baby foods.
Loureiro and Hine	2001	Estimated the willingness to pay for Colorado potatoes.

Academic researchers have taken a slightly different tack in studying organic consumers by complementing surveys with statistical analysis to depict the typical consumer. Some studies reveal conflicting results, most likely because of the different methodological approaches. Consumers considered the following factors important when purchasing fresh produce: price, size and packaging, whether the item is on sale, and whether the item is organic (Estes and Smith). Age, gender, and having a college degree had little impact on a shopper's decision to buy organic produce (Thompson and Kidwell). Consumers with higher incomes and higher levels of education are willing to pay more for organic potatoes (Loureiro and Hine), while consumers with advanced degrees are less likely to buy organic produce (Thompson and Kidwell). Appearance of fresh produce mattered, and the larger the number of cosmetic defects, the less likely would an organic product be purchased (Estes and Smith, Thompson and Kidwell).

Organic Standards and Certification

Organic farming systems rely on ecologically based practices, such as biological pest management and composting; virtually exclude the use of synthetic chemicals, antibiotics, and hormones in crop production; and prohibit the use of antibiotics and hormones in livestock production. Under organic farming systems, the fundamental components and natural processes of ecosystems— such as soil organism activities, nutrient cycling, and species distribution and competition—are used as farm management tools. For example, food and shelter are provided for the predators and parasites of crop pests, planting and harvesting dates are carefully planned and crops are rotated, and animal manure and crop residues are cycled in organic production systems. Organic livestock production systems attempt to accommodate an animal's natural nutritional and behavioral requirements, requiring dairy cows and other ruminants, for example, to have access to pasture.

Private organizations, mostly nonprofit, began developing certification standards in the early 1970's as a way to support organic farming and thwart consumer fraud. Some States began offering organic certification services in the late 1980's for similar reasons. The resulting patchwork of standards in the various certification programs, however, caused a variety of marketing problems. Congress passed the Organic Foods Production Act of 1990 to establish national standards for organically produced commodities, and USDA promulgated final rules for implementing this legislation in December 2000. USDA is currently implementing these organic regulations, and all agricultural products that are sold, labeled, or represented as organic must be in compliance with the regulations after the 18-month transition period is completed in October 2002. These regulations require that organic growers and handlers (including food processors) be certified by a State or private agency accredited under the uniform standards developed by USDA, unless the farmers and handlers sell less than $5,000 a year in organic agricultural products. Retail food establishments that sell organically produced agricultural products but do not process them are also exempt from certification.

The national organic standards address the methods, practices, and substances used in producing and handling crops, livestock, and processed agricultural products. Although specific practices and materials used by organic operations may vary, the standards require every aspect of organic production and handling to comply with the provisions of the Organic Foods Production Act. Organically produced food cannot be produced using genetic engineering and other excluded methods, sewage sludge, or ionizing radiation. These standards include a national list of approved synthetic, and prohibited nonsynthetic, substances for use in organic production and handling.

USDA organic standards for food handlers require that all nonagricultural ingredients, whether synthetic or nonsynthetic, be included on the national list. Handlers must prevent the commingling of organic with nonorganic products and protect organic products from contact with prohibited substances. In a processed product labeled as "organic," all agricultural ingredients must be organically produced unless the ingredient(s) is(are) not commercially available in organic form.

The labeling requirements under the national standards apply to raw, fresh, and processed products that contain organic ingredients and are based on the percentage of organic ingredients in a product. Agricultural products labeled "100 percent organic" must contain (excluding water and salt) only organically produced ingredients. Products labeled "organic" must consist of at least 95-percent organically produced ingredients. Products labeled "made with organic ingredients" must contain at least 70-percent organic ingredients. Products with less than 70-percent organic ingredients cannot use the term organic anywhere on the principal display panel but may identify the specific ingredients that are organically produced on the ingredients statement on the information panel. The USDA organic seal—the words "USDA organic" inside a circle—may be used on agricultural products that are "100 percent organic" or "organic." A civil penalty of up to $10,000 per violation can be levied on any person who knowingly sells or labels as organic a product that is not produced and handled in accordance with these regulations.

For further information, visit USDA's Agricultural Marketing Service/ National Organic

One picture of the typical organic shopper is a younger household in which females do the shopping; smaller and higher income households are the most likely purchasers of organic produce (Govindasamy and Italia) and organic apples (Loureiro et al.). Households knowledgeable about alternative agriculture are more likely to purchase organic produce (Govindasamy and Italia) and those concerned about the environment are more likely to purchase organic apples (Loureiro et al.). Those concerned about food safety are more likely to buy organic produce (Govindasamy and Italia) and organic apples (Loureiro et al.). Those who enjoy trying new products are more likely to purchase organic produce (Govindasamy and Italia). Households with children under 18 are more likely to purchase organic produce (Thompson and Kidwell) and organic apples (Loureiro et al.). Consumers with children are willing to pay less for organic potatoes (Loureiro and Hine) and more likely than other households to purchase organic apples (Loureiro et al.).

Strategies to increase purchases of organic food include shelf-labeling, which had a mixed effect on sales in an upscale grocery store but a significantly positive effect on sales of dairy products, pasta, bread, cereal, and carrots in a discount retailer in the Minneapolis and St. Paul markets (Reicks, Splett, and Fishman).

The recent addition of organic food sales to scanner data, by AC Neilson and Information Resources, Inc., has made possible econometric studies of consumer demand for organic food. Frozen organic vegetables, organic milk, and organic baby food all exhibited high price elasticity of demand, meaning that the quantity purchased responds greatly to price changes (that is, quantity purchased increases by more than 1 percent when prices fall by 1 percent) (Glaser and Thompson, 1999, 2000; Thompson and Glaser, 2001). For some frozen vegetables, there was little crossover between purchases of organic and conventional products, so that changes in prices of either commodity had no significant impact on quantities purchased (Glaser and Thompson 1999). For other products (milk and baby food), the conventional and organic products are substitutes, so that increases in the price of the conventional product result in consumers' purchasing a greater quantity of the organic products (Glaser and Thompson, 2000; Thompson and Glaser, 2001).

The Marketing Chain: from Farm to Market

Food passes through many hands as it moves from farm to consumer. Some foods are fresh when delivered (apples and eggs) while others are processed before delivery (pasta and bread). Regardless of whether they are fresh or processed, higher quality products and products with unique attributes (such as organic foods) generally have a higher selling price. As a result, farmers have a strong incentive to produce and sell commodities with quality and other price-enhancing attributes intact. Yet, since most foods pass through a number of intermediaries as they move from the farm to the consumer, maintaining premium product integrity along the marketing chain can sometimes be a challenge. To do so, each agent along the marketing chain must begin by moving the product to the next agent quickly. Farmers need to sell their perishable commodities immediately after harvesting, while distributors, brokers, and wholesalers need to get fresh products to retailers as quickly as possible. Retailers want to be able to purchase a consistent and large enough supply of a wide variety of uniform quality fresh food. Consumers want to be able to buy a wide variety of fresh food that is both high quality and low priced. Organic food consumers, in particular, want to feel confident that they are buying food that not only was grown organically, but also has kept its organic integrity at each stage in its journey to the market.

Each commodity, depending in large part on whether it is fresh or processed, follows an individualized path from farm to market. Because fresh foods rapidly deteriorate, they must be delivered to the market quickly. The storage and transportation systems along the way must provide the proper temperature and other conditions that help maintain freshness. Processed foods, on the other hand, have a longer shelf life—but the products that go into them must be harvested at the right time, delivered at the right time and satisfy the processor's quality requirements. In the next sections, we trace the production and marketing chains for the major organic commodities in the United States, noting applicable regulations as well as observed marketing trends.

Organic Fresh Fruits and Vegetables

The first stage in the organic fresh fruit and vegetable marketing chain—the production and preparation of produce for shipment— involves growers, packers and shippers, working together in a number of possible combinations. In some cases, one firm grows, packs, and ships the produce, while in others one firm grows and another packs and ships. After it is shipped, produce can either be sold to retailers by a broker or delivered to a terminal market, where it is sold to retailers by wholesalers. In practice, most organic produce is sold through a specialty broker rather than in a terminal market (Dimitri and Richman). In some instances, when a specific variety, quality or quantity is desired, larger retailers may buy fresh fruits and vegetables directly from the produce shipper. Organic produce can also be sold directly to consumers through farmers markets, community- supported agriculture, and farm stands.

Fruit and Vegetable Marketing Chains

Fresh produce:

Farm – shipper – wholesaler – natural foods retailer
Farm – shipper – wholesaler – conventional retailer
Farm – shipper – specialty broker – retailer
Farm – shipper – natural foods retailer
Farm – shipper – conventional retailer
Farm – consumer farmers markets, roadside stands, U-pick, community supported agriculture

Credit: USDA

What are Organic Fruits and Vegetables?

Organic fruit and vegetable production relies on ecologically based practices, such as biological pest management and composting, and crops are produced on land that has had no prohibited substances applied to it for at least 3 years prior to harvest. Soil fertility and crop nutrients are managed through tillage and cultivation practices, crop rotations, and cover crops, supplemented with manure and crop waste material and allowed synthetic substances. Crop pests, weeds, and diseases are controlled through physical, mechanical, and biological control management methods.

Organic fruits and vegetables must be stored and shipped separate from conventionally grown produce. Organic produce is shipped or packed in containers free from synthetic fungicide, preservative, or fumigant.

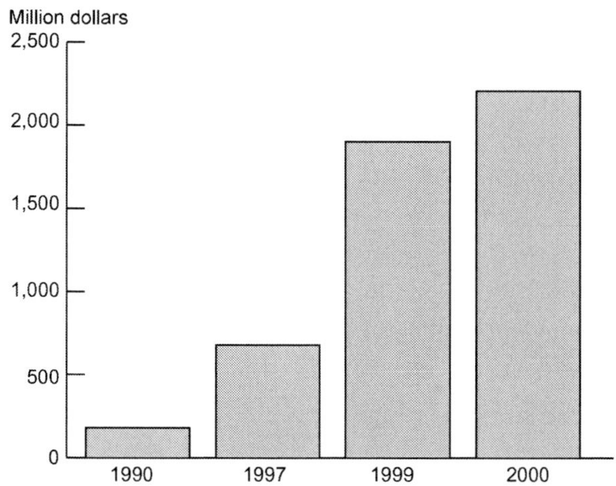

Source: 1997 data. Food Industrial Management Program at Cornell University; 1999, 2000, *Nutrition Business Journal.*

Figure 3. Sales of organic fruits and vegetables: 1990, 1997, 1999, and 2000

Certified Organic Acreage in 1997

California had the most certified organic vegetable acreage, 22,886 acres in 1997. Colorado and Washington followed with 3,716 and 3,140 acres certified in 1997. Arizona, Oregon, Minnesota, New York, Illinois, and Florida each had over 1,000 acres of certified organic vegetables.

Almost 2 percent of U.S. lettuce acreage, 2.4 percent of carrot acreage, and 1 percent of tomato acreage was certified organic.

Over 49,000 acres of organic fruit and nut crops were certified in 36 States. Organic grapes accounted for 39 percent of the acreage certified, followed by apples (18 percent), citrus (12 percent), and tree nuts (10 percent);21 percent of the total was unclassified as to the crop grown.

Growers in California had 32,582 acres of certified organic fruit and nut acreage, over two-thirds of the total. Arizona was second (4,361 acres), and Washington third (2,978 acres).

Fruit and Vegetable Market Facts

Organic produce accounts for 42 percent of all organic food sales (Packaged Facts, 2001).

According to Spins, fresh produce (UPC coded) is the organic category with the third largest level of sales in conventional supermarkets, with sales equal to $93,978,272 in the 12-month period ending June 2001.

Sales of fresh organic produce in natural foods supermarkets totaled $708 million in 1998.Fresh organic produce sales (UPC and non UPC coded) summed to $833 million in 1999, which was 69.4 percent of total fresh produce sales in natural foods supermarkets. Fresh produce was the top-selling organic food category (NFM).

According to Fresh Trends survey, the organic products most often purchased are tomatoes, leafy vegetables, carrots, and apples.

According to the Hartman survey, the organic products most often purchased are strawberries, lettuce, and carrots.

Fruits and vegetables are the "gateway" category into organic foods (Hartman Group, 2001).

Source: Greene

Price Premiums for Organic Produce in the Boston Wholesale Market During 2000 - 2001

Organic broccoli – 30 percent higher, on average, than conventional prices
Organic carrot – 25 percent higher, on average, than conventional prices
Organic mesclun – 10 percent higher, on average, than conventional prices

Source: Sok and Glaser.

Credit: PhotoDisc

ORGANIC GRAINS, OILSEEDS, AND LEGUMES

Organic grains, legumes are used as inputs to oilseeds, and manufactured products, as feed grain and as final food products such as rice and tofu. Crops produced by organic grain and oilseed farmers include traditional grains and oilseeds such as corn, soybeans, wheat, barley, oats, and rice, as well as nontraditional grains, including millet, buckwheat, rye, and spelt.

Organic grains, oilseeds, and legumes may be contracted prior to planting. Several companies offer producers contracts for organic wheat, corn, soybean, black bean, buckwheat, sunflower, popcorn, spelt, and millet crops. Contracts for organic grains vary considerably— for example, some make producers responsible for grain cleaning or shipping charges or both while others do not make the producer responsible for either (Born and Sullivan).

Marketing Chain for Organic Grains, Oilseeds, and Legumes

Farmer – cooperative – cleaner – manufacturer – distributor
Farmer – cleaner – manufacturer – distributor
Farmer – cooperative – cleaner – broker – manufacturer – distributor

Farmer – cleaner – broker – manufacturer – distributor
Farmer – marketing agent (often contracts with farmer, and cleans) – manufacturer
Farmer – cooperative – processor of feed grain – distributor – livestock producer
Farmer – processor of feed grain – distributor – livestock producer

Credit: Comstock Images.

Top-Selling Products Made with Organic Grains and Legumes

Soy milk and rice milk: the organic nondairy beverage category had sales in all venues of $316 million in 1999 and $395 million in 2000 (NBJ).

Organic nondairy beverages made up 62 percent of U.S. sales of nondairy beverages in all venues in 1999 (NBJ).

Sales of organic breads and grains increased from $700 million in 1999 to $848 million in 2000 (NBJ).

Sales of organic tofu totaled $40.8 million in 1999 and $46.3 million in 2000. Slightly more than half of organic tofu sales (52.8 percent) were made in conventional supermarkets (NFM, 2001).

What are Organic Grains, Oilseeds, and Legumes?

Organic grain, oilseed, and legume production relies on ecologically based practices, such as biological pest management and composting, and crops are produced on land that has had no prohibited substances applied to it for at least 3 years prior to harvest. Soil fertility and crop nutrients are managed through tillage and cultivation practices, crop rotations, and cover crops, supplemented with manure and crop waste material and allowed synthetic substances. Crop pests, weeds, and diseases are controlled through physical, mechanical, and biological control management methods.

When being stored and shipped, organic grains, oilseeds, and legumes must be kept from commingling with conventionally grown grain. These organic products cannot be shipped or packed in containers containing synthetic fungicide, preservative, or fumigant.

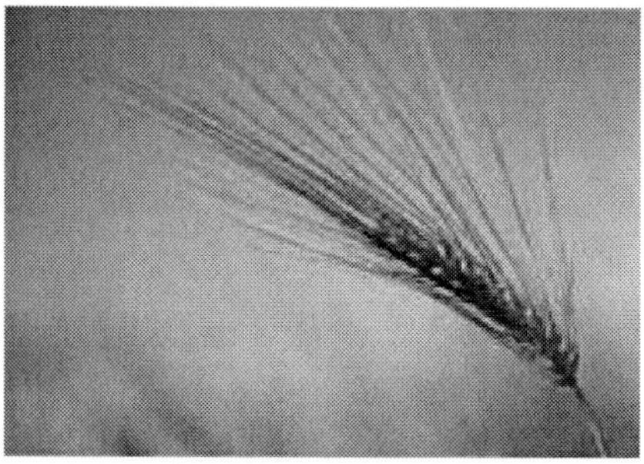

***Credit**: PhotoDisc.*

Organic Soybean and Legume Market Facts

Soybeans are the top legume crop grown under certified organic farming systems.U.S. growers produced over 82,000 acres of certified organic soybeans in 1997. Organic soybeans accounted for 0.1 percent of total soybean acreage in the United States (Greene).

Certified organic dry peas and lentils were grown on nearly 5,200 acres in the United States in 1997.Mon-

Price premiums for organic grains and oilseeds for the time period between 1995 and 2000

	1995	1996	1997	1998	1999	2000	2001
				Percent			
Corn	35	43	73	88	98	89	59
Soybeans	114	85	141	202	217	175	177
Spring wheat	54	59	73	8	87	103	94
Oats	35	59	73	83	77	71	41

Note: The premiums are reported as the percent higher than prices for the conventionally produced equivalent.

Source: Bertramsen and Dobbs, 2002.

PROCESSED ORGANIC FOODS

Organic processed foods include frozen vegetables and entrees, pasta, canned vegetables, sauces in jars, and shelf-stable entrees. Specific products must be used to manufacture these foods—for example, pasta processors need to use a particular variety and grade of organic wheat, while frozen fruit and vegetable processors need organic produce of a specific size and quality. All processors want uniform quality, so they can offer products that consistently taste the same. Consequently, the biggest challenge facing organic manufacturers is how to secure a steady supply of organic ingredients of a consistent quality. Often, manufacturers will recruit new farmers, assisting them with converting from conventional to organic production. Retailers with private label organic products may also provide on-farm assistance to producers. The next biggest challenges are how to transport the processed goods to the supermarket and how to secure shelf space.

There are two basic marketing channels for processed foods. In both cases, farmers first produce raw commodities. In the first case, these commodities are then sent to the manufacturer, who converts them into a processed product. A distributor acts as a middleman, moving processed products from manufacturers to retailers. In the second scenario, a middleman (shipper) procures raw commodities from farmers and delivers them to manufacturers. After creating the processed good, the manufacturer moves the products through to retailers. The middleman secures the quantities needed; he or she also ensures that the commodities are high quality and meet the manufacturer's organic standards.

Photo courtesy of Organic Valley Family™ of Farms, 2002.

Processed Products Marketing Chain

Farm – manufacturer – wholesaler – retailer
Farm – shipper/procurer – manufacturer – wholesaler – retailer

Average price premiums for organic frozen vegetables, 1991-1996

Frozen broccoli	72 percent
Frozen sweet corn	25.8 percent
Frozen green peas	109.7 percent
Frozen green beans	75.7 percent

Note: The premiums are reported as the percent higher than prices for the conventional equivalent.
Source: Glaser and Thompson, 1999.

What are Organic Processed Products?

A certified organic processed product, such as pasta or frozen pizza, is first prepared using at least 95 percent organic ingredients. Conventionally grown ingredients may be used only when an organic ingredient is not commercially available. For products that contain 70 to 95 percent organic ingredients, processors may label the organic ingredients on the package.

Organic and conventionally grown ingredients must be kept separate, and the organic ingredients must be stored in containers that do not compromise the organic nature of the food. Both organic and conventional ingredients must not be treated with ionizing radiation, excluded methods, and synthetic solvents.

Select manufactured organic food sales by venue

Store	Entrees	Desserts	Frozen foods	Canned & jarred*	Pasta	Cereal	Oils	Soy
				Percent				
Health/natural foods stores	42	82	47	68	59	64	75	21
Conventional stores	55	10	47	30	41	31	11	77

*The category includes soups, canned fruits and vegetables, and baby food.
Note: The percentages do not sum to 100 because manufacturers sell in other venues, such as food service and club stores, which are not listed.
Source: OTA 2001 Manufacturers' Survey; for products that contain at least 50 percent organic ingredients.

Where are Organic Manufactured Foods Sold?

In 2001, 45 percent of organic manufactured products were sold in conventional grocery stores compared with 31 percent in 1998 (OTA Manufacturers' Survey, 2001).

In 2000, 51.3 percent of frozen organic entrees, pizza, and convenience foods were sold in conventional supermarkets, which increased to 53.1 percent in 2001 (NFM, 2002, 2001).

Credit: PhotoDisc.

The dollar amount of frozen organic entrees, pizza, and convenience foods sold in conventional stores, in the 12- month period ending June 2001, was $74 million (Spins, as quoted in Organic and Natural Business News).The dollar amount sold in all venues in 2001 totaled $152 million (NFM, 2002).

Organic packaged grocery sales in natural products stores totaled $652 million in 2000 and increased to $1.6 billion in 2001. Organic frozen/refrigerated product sales totaled $333 million in 2000 and increased to $823 million in 2001 (NFM, 2001).

In 2001, 49 percent of manufactured organic foods were sold in health and natural foods stores, and 45 percent in conventional grocery channels.

In 1998, 62 percent of manufactured organic foods were sold in health and natural foods stores; 31 percent through conventional grocery channels (OTA,1998).

Organic Processed Foods Market Facts

Frozen organic fruit and vegetable sales as a share of total U.S.frozen fruit and vegetable sales:1.34 percent.

Canned organic fruit as share of total U.S.canned fruit: 0.63 percent.

Canned organic vegetables as share of total U.S.canned vegetables:0.43 percent.
Source: *Nutrition Business Journal*

In 2000, frozen prepared foods were among the fastest growing processed foods categories.

Packaged groceries accounted for 15 percent of total organic sales.

Frozen foods accounted for 8 percent of organic sales.
Source: Packaged Facts

Sales of four types of frozen organic vegetables (corn, broccoli, peas, and green beans) increased by 58 percent per year between 1991 and 1996 (Glaser and Thompson, 1999).

In the first half of 2000, more than 800 organic new processed products were introduced; the number of dessert items increased the most (Meyers and Rorie).

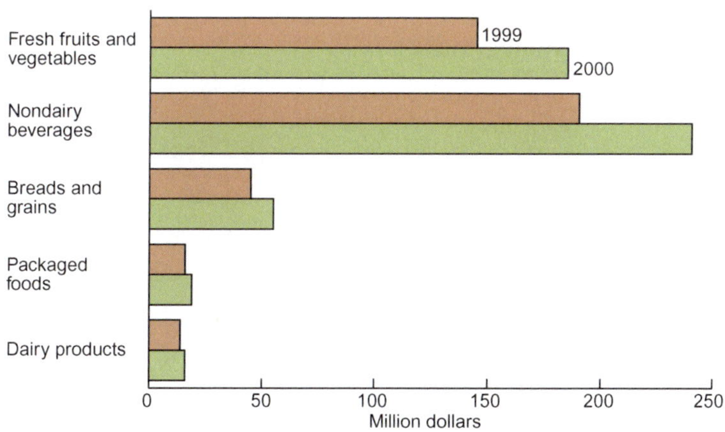

Source: Nutrition Business Journal.

Figure 4. Sales of select processed organic foods

Organic Baby Food

In 2000, 73.9 percent of organic baby food was sold in conventional supermarkets, and 21 percent was sold in natural foods stores (NFM, 2001). In the

12-month period ending June 2001, sales of organic baby food in conventional supermarkets totaled $50.6 million.

Organic baby food sales comprised 4.5 percent of all baby food sales in 1999. (NBJ, 2001).

Price premiums for organic baby food sold in conventional supermarkets ranged from 35 to 53 percent in 1999 (Thompson and Glaser, 2001

ORGANIC DAIRY PRODUCTS

Organic dairy products, as defined by the USDA, are made from the milk of animals raised under organic management. The cows are raised in a herd separate from conventional dairy cows. The animals are not given growth hormones or antibiotics. The animals receive preventive medical care, such as vaccines, and dietary supplements of vitamins and minerals. All organically raised dairy cows must have access to pasture, the outdoors, shade, shelter, exercise areas, fresh air, and direct sunlight suitable to their stage of production, the climate, and the environment.

To convert from conventional to organic production, the cows must be fed a diet consisting of at least 80 percent organic feed for 9 months and then 100 percent organic feed for 3 additional months, or must be grazed on land that is managed under a certified organic plan. Otherwise, organic dairy products must make use of milk from animals raised organically for at least 1 year prior to producing the milk.

Photo courtesy of Organic Valley Family™ of Farms, 2002.

The process used to bottle milk and to make and pack cheese, ice cream, yogurt, and other dairy products must also be certified. The processor is required to keep organic and nonorganic products separated, and must prevent organic products from contact with prohibited substances. Nonorganically produced ingredients may be used when processing an organic product if there is no organic equivalent. Volatile synthetic solvents and other synthetic processing aids are prohibited.

Organic manufacturers'2001 sales of dairy products were distributed as follows

Health and natural food stores:65% cheese, 50% yogurt, butter, cheese, sour cream and kefir;33% milk and cream

Conventional supermarkets:33% cheese;47% yogurt, butter, cheese, sour cream and kefir; 65% milk and cream.

Source: OTA's 2001 Manufacturers'Survey.

Sales of organic dairy products, in millions of dollars

	1999	2000
Milk, cheese, butter, eggs, yogurt	394	480
Frozen dairy, ice cream	85	106
Canned milk	2	2

Marketing Channels

Milk (from one or several farms) – on-farm dairy (pasteurize and bottle) – regional distributors

Milk (from one or several farms) – off-farm dairy (pasteurize and bottle) – regional distributors

Milk (from one or several farms) – off-farm processors of cheese, butter, yogurt, or dry milk – distribute regionally and nationally

Milk (from several farms) – dairy (pasteurize and bottle, or process into cheese or ice cream, for example) – distribute nationwide through a marketing cooperative (under a brand name), mainly through private distribution networks, sometimes with a regional label.

Milk produced under contract (from several farms) – dairy (pasteurize and bottle, or process into cheese or ice cream, for example) – distribute nationwide under a brand name, mainly through private distribution networks

Milk Market Facts

Organic milk first appeared in conventional supermarkets in 1993; 8 conventional supermarkets sold organic milk in December 1996 (Glaser and Thompson, 2000)

Organic milk, half and half, and cream sales in conventional markets:

1996: $15.8 million; (Glaser and Thompson, 2000)
1997: $30.7 million (NFM).
1998: $46.0 million (Glaser and Thompson, 2000)
1999: $75.7 million (NFM)
2000: $104 million (NFM)

Organic milk, cream, half and half sales in natural products supermarkets:

1999: $40 million (NFM)
2000: $55 million (NFM)

Price premium for organic milk over branded or private label conventional products was in the range of 50 – 72 percent between 1996 and 1999 (Glaser and Thompson, 2000).

The price premium for organic milk ranged from 47 percent to 76 percent in the Twin Cities region of Minnesota between 1995 and 1997 (Dobbs et al.).

The price premium for organic cheese ranged from 124 percent (1995) to 28 percent (1997) in the Twin Cities region of Minnesota (Dobbs et al.).

Organic dairy products (milk, butter, cheese, eggs, yogurt) comprised 0.9 percent of total U.S. dairy sales in 2000 (Nutrition Business Journal).

Photo courtesy of Organic Valley Family™ of Farms, 2002.

In 1997, there were 12,897 certified organic dairy cows. New York was the top producer (3,386 dairy cows), followed by Wisconsin (2,509) and Minnesota (2,425). Pennsylvania, California, and Maine also had over 1,000 organic dairy cows each (Greene).

The number of certified organic milk cows nearly tripled between 1992 and 1994 and more than doubled between 1994 and 1997 (Greene).

Share of dairy product sales in natural foods supermarkets and conventional grocery stores

In conventional supermarkets, milk, half & half, and cream is the second leading category of organic food sales, with sales equal to $119,315,772 in the 12-month period ending June 2001*

Product	**1999**		**2000**		**2001**	
	Natural	**Conventional**	**Natural**	**Conventional**	**Natural**	**Conventional**
			Percent			
Milk, half and half, cream	35.1	64.6	34.6	65.3	30.53	69.25
Cheese and cheese alternatives	85.4	14.6	46.3	51.7	43.71	55.97
Yogurt and kefir	56.8	43.2	52.4	47.4	43.17	56.70

*SPINS, as quoted in *Organic and Natural News,* 2001.

Source: *Natural Foods Merchandiser*, 2000, 2001, and 2002.

ORGANIC MEAT, POULTRY, AND EGGS

Producing organic beef has three phases. The first is cow-calf, which is the period from birth to weaning, when the calf weighs approximately 500 pounds. The second phase is back-grounding, or the period between weaning and the time the calf weighs 900 pounds. The finishing phase, done in a feedlot for conventionally produced beef, is a 3- month period prior to slaughter, by which time the calf weighs between 800 and 1,200 pounds. During the cow-calf and backgrounding phases, the animals must be fed organically grown pasture and hay; some farmers choose to add organically grown grain to the cattle's diet during the last months prior to slaughter. Organic farmers must maintain organic pasture for the cattle to graze on throughout all three production stages (thus, the term "grass-fed"). The farmer may produce the supplemental feed, such as grain or hay, may grow a portion and purchase some, or may purchase all the feed. Organic grass-fed beef production, from birth to slaughter, takes approximately 26 – 28 months.

Organic hog production begins with the gestating sow, right before giving birth (farrowing), building a cozy place to birth her litter, where the sow and her litter will live for 1 week after birth. After nursing for 5 weeks, the pigs are weaned and then pastured. Pigs not raised on pasture may live in indoor or indoor/outdoor housing. Since hogs eat their bedding, it must be organically produced. Bedding materials might include small-grain straw, corn stalks, soybean straw, or grass hay (Gegner). Pasture-raised hogs eat grass, legumes, standing crops and other ground cover (Gegner). The farmer may produce the supplemental feed, such as grain or hay, may grow a portion and purchase some, or may purchase all the feed. The hogs reach market weight (240 – 260 pounds) in 6 to 6.5 months from birth.

Credit: USDA.

Organic poultry (chicken and turkey) are raised organically, beginning from at least the second day of life. Some farmers purchase chicks from a certified organic hatchery while others begin raising the chicks organically when they arrive on the farm. The birds eat organically produced grain, and are market weight in 70 to 81 days. The farmer may produce the supplemental feed, such as grain or hay, may grow a portion and purchase some, or may purchase all the feed.

Organic Pork & Poultry Marketing Channels

National Distribution

Farm – cooperative (also processes) – retailer
Farm – processor – distributor – retailer

Local Distribution

Farm – consumer pickup on farm
Farm – consumer via farmers market
Farm – consumer via internet sales

Photo courtesy of Organic Valley Family™ of Farms, 2002.

Organic Beef Production and Marketing Channels

Nationwide Distribution

Cow/calf farm – pasture farm – cooperative (also processes) – retailer
Cow/calf & pasture farm – cooperative (also processes) – retailer
Cow/calf farm – pasture farm – farm – processor – distributor – retailer
Cow/calf & pasture farm – farm – processor – distributor – retailer

Local Distribution

Cow/calf farm – pasture farm – consumer

Cow/calf & pasture farm – consumer

Nationwide distribution often relies on a brand name or marketing cooperative.

Several small, often family-run, farms raise livestock or poultry organically, which are processed by the cooperative or firm holding the brand name. The meat products are distributed to retailers by the cooperative or a distribution firm.

Organic Livestock, Poultry, and Egg Market Facts

Eleven percent of the meat sold in natural products stores in 2000 was organic (NFM).

Sales of organic eggs in natural foods stores and conventional supermarkets totaled $37.3 million in 1999, and $42.3 million in 2000 (NFM, 2001). In 2000, consumers purchased 64.9 percent of all organic eggs in conventional supermarkets, 35 percent in natural foods stores, and 0.1 percent direct from the producer (NFM, 2001).

Organic meat comprised 3 percent of total organic sales in 2000 (Myers and Rorie). According to the OTA Manufacturers' Survey, members reported that sales of meat and poultry products increased 151 percent between 1999 and 2000. Members also reported that 78 percent of organic meat was sold in natural foods stores and 15 percent was sold in conventional supermarkets (OTA 2001).

Certified organic livestock is well under 1 percent of total U.S. livestock. However, the number of dairy cows and layer hens produced under certified organic farming systems increased sharply during the 1990s, and the market for organic meat products is expected to grow now that organic labeling is permitted (Greene).

Farmers and ranchers raised certified organic cows, pigs, and sheep in 23 States in 1997 (table 15). Dairy cows led, with 12,897 animals certified organic in 1997. New York was the top producer (3,386 dairy cows), followed by Wisconsin (2,509) and Minnesota (2,425). Pennsylvania, California, and Maine also had over 1,000 organic dairy cows each (Greene).

Thirteen States contained 4,429 certified organic beef cows in 1997, and almost 40 percent of which were in Michigan. Only 482 hogs and pigs were certified organic in 1997 (Greene).

The United States had 537,826 certified organic layer hens in 1997, along with 38,285 organic broiler hens, 750 turkeys, and 221,389 unclassified organic poultry animals. California was the leader in organic poultry production, with 350,000 organic birds, followed by New York (161,304) and Pennsylvania (66,300 birds) (Greene).

Photos courtesy of Organic Valley Family™ of Farms, 2002.

What are Organic Meat, Poultry, and Eggs?

Organic meat, poultry, and eggs, as defined by USDA, are made from animals raised under organic management. All organically raised herds and flocks must be raised separate from their conventional counterparts. The animals are not given growth-producing hormones or antibiotics. The animals receive preventive medical care, such as vaccines, and dietary supplements of vitamins and minerals. They consume 100 percent organically produced feed, free of animal byproducts. Producers must provide living conditions that accommodate the health and natural behavior of the animals. The animals should have access to the outdoors, shade, exercise areas, fresh air, and direct sunlight suitable to their species and stage of production. All organically raised cows must have access to pasture. Their bedding must be clean and dry. Producers may not withhold medical treatment from a sick animal in order to preserve its organic status. The producer must manage manure in a way that does not contribute to soil, water, or crop contamination. In certain cases, the producer may temporarily confine animals because of weather, stage of development, if the animal's well-being would be compromised from being outdoors, or if there was a risk to soil or water quality.

Livestock intended for meat products must be raised organically from the last third of gestation. Livestock used as breeder stock can be brought from a conventional operation provided that the animal is raised organically for the last third of gestation, and the offspring is raised organically from birth.

Poultry and layer hens must be under continuous organic management from the second day of life (most producers start raising poultry from day-old chicks). The recommended method of farming is for poultry to have access to the outdoors during the months when it is feasible. In order to become certified, the producer must demonstrate how access to the outdoors will be encouraged and maximized.

Organically produced feed grain is created in certified organic mills. Cows, hogs, and poultry are processed in certified organic plants.

ORGANIC FIBERS

Several types of fiber are organically produced, including cotton, rayon, and linen. Certified organic cotton was produced in 4 States—Texas, New Mexico, Missouri, and California—on 9,974 acres in 1997. Texas had over three-fourths of the U.S. organic cotton acreage in 1997, much of it operated by an organic cotton marketing cooperative that formed in the early 1990s and is still active. Certified organic cotton acreage in 1997 was 70 percent lower than in 1995, and accounted for only 0.1 percent of U.S. cotton acreage. Currently, the organic cotton market is more developed than markets for other organic fibers.

Several major clothing companies tested organic cotton clothing lines without commercial success in the mid-1990s. These companies have now switched to a new approach—blending organic with conventional cotton—to help stabilize the market and encourage organic production (Bunin, 2000).

Acres of certified organic and transitional land and production by State

	1997	2001	2000 production
	Acres	*Acres*	*Bales*
Texas	8,134	8,338	3,128
Arizona	0	1,303	2,122
California	756	828	1,630
New Mexico	512	470	1,146
Missouri	572	400	510
Kansas	0	70	5
Tennessee	0	50	0

Source:1997, Greene;2000 and 2001, Organic Trade Association.

What are Organic Fibers?

Organic fiber production relies on ecologically based practices, such as biological pest management and composting, and crops are produced on land that has had no prohibited substances applied to it for at least 3 years prior to harvest.

Soil fertility and crop nutrients are managed through tillage and cultivation practices, crop rotations, and cover crops, supplemented with manure and crop waste material and allowed synthetic substances. Crop pests, weeds, and diseases are controlled through physical, mechanical, and biological control management methods.

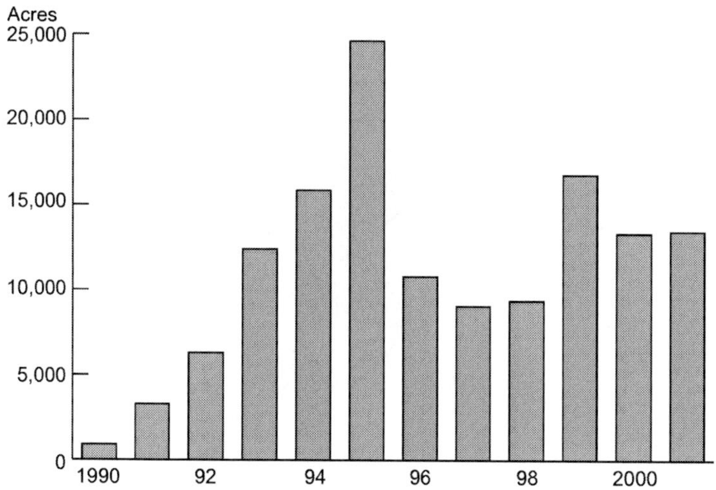

Source: Organic Trade Association.

Figure 5. Organic cotton acreage, 1990 to 2001

Credit: Digital Stock

Organic Herbs and Flowers

Most organic herbs and flowers (excluding those used as an intermediate good) are sold locally and direct to the consumer, through internet sales, CSAs, or farmers markets. Natural products supermarkets throughout the United States sell organic herbs (most of which are grown locally). Natural products supermarkets in the Pacific Northwest have just begun selling organic flowers.

Credit: USDA.

What are Organic Herbs and Flowers?

Organic herb and flower production relies on ecologically based practices, such as biological pest management and composting, and crops are produced on land that has had no prohibited substances applied to it for at least 3 years prior to harvest. Soil fertility and crop nutrients are managed through tillage and cultivation practices, crop rotations, and cover crops, supplemented with manure and crop waste material and allowed synthetic substances. Crop pests, weeds, and diseases are controlled through physical, mechanical, and biological control management methods.

Marketing Chain for Organic Herbs and Flowers

For herbs as a final product and flowers:

Producer – consumer

Producer – processor – wholesaler – retailer

For herbs used an intermediate good:

Producer – processor – manufacturer

Organic Herb and Flower Market Facts

Organic herbs are used for culinary purposes, dietary supplements, such as St. John's Wort, and for personal care products, such as body lotion.Those used for culinary reasons are either sold as spices to consumers or are sold in large quantities to manufacturers to use in manufacturing prepared foods (Frontier).

Certified organic herbs were cultivated on 6,407 acres in 1997, wild organic herbs were also harvested from an additional 83,388 uncultivated acres[4] (Greene).

Producers grew certified organic herbs for culinary and medicinal uses in 32 States.California was the largest producer of cultivated organic herbs, with 1,062 acres, followed by Washington (846 acres) and Illinois (797 acres).Seven other States (Oregon, North Carolina, Florida, Missouri, Minnesota, Colorado, and Wisconsin) had over 200 acres of cultivated herbs.Hundreds of different types and varieties are being cultivated in these States (Greene).

Idaho had 52,000 acres of wildcrafted St.John's Wort, a popular medicinal herb, in 1997.Florida had 25,000 acres of certified organic saw palmetto berries (which have medicinal uses) and maypop (which have culinary and medicinal uses), indigo (which is used as a natural hair dye), polk (a wild salad green), and other crops.Oregon certified 6,000 acres of lake algae (which has medicinal uses) and St.John's Wort (Greene).

Certified organic cut flowers were produced in a dozen States on 288 acres in 1997.California was the largest producer with 145 acres (Greene).

Appendix: USDA Research, Program, and Regulatory Activities on Organic Agriculture

Federal Government efforts to facilitate organic production have focused primarily on developing national certification standards to assure consumers that certified organic commodities meet consistent standards. Now, a small number of new programs and pilot projects are underway to help organic producers with production problems and risks, and promote organic agricultural products overseas. The following sections outline the research, program, and regulatory efforts USDA is undertaking for organic agriculture.

USDA's Agricultural Marketing Service and Organic Agriculture

Contributed by Demaris Wilson, USDA-AMS

The Agricultural Marketing Service (AMS) is home to the National Organic Program (NOP), which developed, implemented, and administers national production, handling, and labeling standards for organic agricultural products. The NOP also accredits the certifying agents (foreign and domestic) who inspect organic production and handling operations to certify that they meet USDA standards. To facilitate the export of U.S. organic agricultural products, the NOP is working to establish formal recognition agreements with foreign governments.

In addition to its regulatory duties,AMS provides information to consumers via the NOP web page *www.ams.usda.gov/nop.*

The AMS Fruit and Vegetable (FV) Market News has provided some market coverage for organically grown fruits and vegetables at a few wholesale markets across the country for a number of years. To determine if additional market coverage could be provided with current resources, FV Market News reporters surveyed their contacts at both shipping point and wholesale markets to determine which markets handle organically grown produce as part of their normal or seasonal product line. A key aspect of the upcoming survey report will assess the willingness of contacts to provide price and volume data of organically produced fruit and vegetables separately from that of conventionally produced produce.

AMS is also involved in several areas of organic marketing research, working independently and in cooperation with major universities. AMS has formed a partnership with the University of California-Davis to study how existing fruit and vegetable marketing orders will be affected by the national organic standards and to determine marketing opportunities for organic commodities, particularly California almonds and kiwifruit and winter pears from the Northwest.

And, finally, to determine the availability of certified organic feed for use in organic livestock operations,AMS has a cooperative agreement with Iowa State University and North Carolina State University to canvass the amount of certified organic acreage that is planted in corn and soybeans.

USDA's Alternative Farming Systems Information Center

National Agricultural Library
Contributed by Mary Gold, USDA-NAL

The Alternative Farming Systems Information Center (AFSIC) is one of several topic-oriented Information Centers at the National Agricultural Library (NAL). The Library, located in Beltsville, Maryland, houses the Nation's largest collection of agricultural resources.

AFSIC serves as an information clearinghouse, specializing in locating and disseminating information related to alternative cropping systems including sustainable, organic, low-input, biodynamic, and regenerative agriculture. AFSIC also focuses on alternative crops, new uses for traditional crops, and crops grown for industrial production. The Center's users are farmers, extension staff, researchers, and educators, as well as consumers and the general public. They access AFSIC's staff and resources from all over the world, via a Web site, *http://www.nal.usda.gov/afsic/*, or by email, phone, fax, and surface mail.

Staff members create publications that focus on specific topics of current interest. These publications are primarily research guides, and include bibliographies, directories and cyberguides. They reflect the materials contained in the National Agricultural Library collection and/or indexed in NAL's literature database,AGRICOLA, as well as electronic data and internet sites worldwide.

AFSIC's Web site currently emphasizes resources related to organic food production *(http://www.nal.usda.gov/afsic/ofp/).* Unique publications include *Tracing the Evolution of Organic/Sustainable Agriculture, Organically Produced Foods: Nutritive Content; Organic Production: Terminology / Descriptive Phrases; Organic Production: Economic Aspects. Books, Articles, and Videocassettes, 1991-March 1997; and Organic Information Resources:What are They? Where are They? How Can I Find Them?* All AFSIC publications are available in full text format at the AFSIC Web site.

The Center has also initiated a project that will identify pre-1945 USDA publications that reflect research pertinent to current organic farming systems and will make selected documents available electronically via its Web site.

USDA's Agricultural Research Service and Organic Agriculture

Contributed by Mike Jawson, USDA-ARS

The Agricultural Research Service (ARS) is USDA's intramural research agency, conducting research at over 100 locations across the United States and at five overseas sites. ARS research addresses all aspects of agriculture from natural resource management to human nutrition. ARS research is organized into 22 national programs, which are described at *www.nps.ars.usda.gov.* There is not a national program on organic agriculture. ARS intends to incorporate organic producers' needs across all the appropriate national programs. The Integrated Farming Systems National Program, however, has been serving as the focal point for many in the sustainable and organic agriculture communities. In the coming months, ARS will post a new Web site featuring its organic farming research.

ARS conducts much research that is applicable to organic agriculture in areas such as soil management, biological control of pests and weeds, crop and animal production and maintenance of germplasm. More than 125 ARS scientists are engaged in research applicable to organic agriculture. Since most of this research is not conducted within a purely organic agriculture context, it requires additional research to test its applicability within organic production systems. ARS is now conducting more and more research in certifiable organic systems, often with on-farm producer involvement. To ensure that organic farming research meets growers' needs, ARS has established a strong working relationship with the Organic Farming Research Foundation.

ARS now has long-term integrated systems projects dedicated to organic agriculture. Projects involving direct producer involvement have been initiated in Maryland, California, Texas, Florida, Iowa, and West Virginia. Other organic projects that build ongoing efforts are under way in Nebraska, Maryland, Minnesota, Washington, Oregon, Wisconsin, and elsewhere. For example:

- Oregon scientists are working with organic growers to quell plant diseases.
- Scientists in California responded to the needs of local organic strawberry growers with some first-ever studies to study commercially available strawberry varieties grown on fields managed organically and are developing organic production systems.
- Researchers in Washington have developed an organic post-harvest treatment for fruit.
- ARS scientists in Florida played a pivotal role in developing a multi-agency group to which Florida's organic growers can turn for help. In current studies, the team is working with local organic growers to enhance weed and disease control on their vegetable farms. They are experimenting with novel cover crops, paper mulches, soil solarization, and biological control agents.
- Beltsville, Maryland, research is investigating organic methods to control animal parasites.

- Grain producers are cooperating with ARS researchers in Maryland, Iowa, Minnesota, and Washington to devise reduced tillage organic systems that provide effective weed control.

USDA's Cooperative State Research,Education,and Extension Service and Organic Agriculture

USDA's Sustainable Agriculture Research and Education (SARE) program
Contributed by Jerry DeWitt, USDA-CSREES SARE

Since 1988, the USDA-CSREES SARE program has helped advance farming systems that are profitable, environmentally sound, and good for communities through an innovative research and education grants program. The information SARE has gleaned from more than 2,100 projects is more relevant than ever, as many research findings have suggested ways to counter low agricultural prices by reducing input costs and expanding market opportunities.

SARE has funded around 391 projects—or 19 percent of its portfolio—that focus on some aspect of organic production and marketing. SARE funds research and education programs through four regional offices in Vermont, Georgia, Utah, and Nebraska.

Most of the SARE grants that support organic agriculture are research and education projects, which involve scientists, producers and others in an interdisciplinary approach to explore new farming systems, methods, and educational opportunities. SARE makes sure that project findings get put to use through its Professional Development Program (PDP), which offers educational opportunities in the latest sustainable practices and systems to extension educators and other agricultural advisers. SARE runs a Producer Grants Program, providing farmers and ranchers an opportunity to test experiments on site and share the results with their peers. SARE also supports the Sustainable Agriculture Network (SAN), which publishes materials for producers and educators. See *www.sare.org* for additional information.

All of SARE's grant programs include opportunities for funding of organic projects. Here is a sampling of some of those:

- A widespread sweet corn pest, corn earworm moths, seek the sweet odor of corn silk to lay their eggs, compelling producers nationwide to accept wormy corn or apply broad-spectrum pesticides 3 to 10 times per crop. Organic growers, in particular, are forced to accept wormy corn. Thanks to work headed by SARE-funded researcher Ruth Hazzard at the University of Massachusetts, growers now use new, effective biological controls to fight the earworm—corn oil and *Bacillus thuringiensis* (Bt). Hazzard's technique calls for applying Bt and corn oil to the top of each ear during the formative stage, causing earworms that crawl down the silks into the ear to suffocate. Eight farmers from Vermont to Connecticut testing the method found that the oil controlled ear damage in 83 percent of their trial plots in 2000. [For more information, see *http://www.sare.org/projects/san_db_viewer.asp?id=1439]*

- When a national organic dairy opened on Maryland's Eastern Shore in 1996, mid-Atlantic grain producers realized they had an opportunity to add value to their product. They knew how to grow corn and beans, but now they wanted to do so organically—and needed help. Recognizing that new niche, University of Maryland extension educator John Hall applied for a SARE grant to create tools that agricultural professionals could use to teach farmers the basics of organic grain production. The final product, a three-part video series, provides essential production information and a colorful mix of examples from successful organic grain farmers. University researchers explain how to create diverse agricultural systems with innate abilities to combat pests, use minimum tillage to minimize compaction and preserve insect habitats, and plant cover crops to build the soil. The project has spawned a nonprofit institute in eastern Maryland that is exploring other marketing outlets for organically produced grain. [For more information, see *http://www.sare.org/projects/san_db_viewer.asp?id=1296]*
- Florida organic farmers seeking an alternative to expensive organic fertilizers tested a new technology that converts food waste to liquid fertilizer. The digester resides at a farm, where about 2 tons of cafeteria food waste is handled each week. Anaerobic bacteria in the digester convert the food waste to methane and carbon dioxide, while the nutrients remain in the water or in a small amount of solid residue that can be applied to land directly or cured to a mature compost. The fertilizer nitrogen value produced from 2 tons of food waste per week ranges from $800 to more than $10,000 annually if calculated on the basis of nitrogen values in poultry litter or fish emulsion, respectively. [For more information, see *http://www.sare.org/projects/san_db_viewer.asp?id=1160]*

"Organic Transitions"and Other CSREES Competitive Grant Programs
Contributed by Tom Bewick, USDA-CSREES

CSREES recently initiated an "Organic Transitions" competitive grants program on pest management to assist farmers in adopting organic practices. This program supports systems research on organic farming combined with outreach and education programs to help farmers apply the results of that research (see *http://www.reeusda.gov/agsys/pestmgt/organic.htm*). The maximum award is for up to 4 years, with no funding limit set.

Three collaborative farmer-researcher projects on organic weed management—on citrus in Florida, organic vegetables in southern California, and limited-resource and family farms in Iowa—were funded in FY 2001. FY 2002 projects include: (1) weed management programs that strengthen the systems approach of organic agriculture, including the effects of soil biology, cover crops, crop rotations, crop/livestock integration and grazing, on weed severity and impact; (2) understanding the relationship of applied organic fertility management to crop health and the resistance of crops to pests and diseases; and (3) designing training systems to elevate the awareness of county Cooperative Extension personnel and other farm advisors about organic practices and information on a national or regional level with particular emphasis on weed management, insect pest management, soil fertility enhancement, best organic cultural practices and livestock management. Priorities for funding in FY 2002 were based, in part, on the biennial survey of research needs of organic farmers conducted by the Organic Farming Research Foundation. A summary of the most recent survey is available online at *http://www.ofrf.org/publications/survey/index.html.*

Also, the 2002 Farm Act authorizes new mandatory appropriations for fiscal years 2003-07 under the Organic Agriculture Research and Extension Initiative. CSREES will administer a competitive research grants program with these new funds. Research is to focus on determining desirable traits for organic commodities; identifying marketing and policy constraints on the expansion of organic agriculture; and conducting advanced research on organic farms, including production, marketing, and socioeconomic research.

USDA's Economic Research Service and Organic Agriculture

The Economic Research Service (ERS) conducts economic research and develops and distributes a broad range of economic and other social science information and analysis on organic agriculture. ERS provides information and analysis on organic farming and marketing in the organic briefing room available online at *http://www.ers.usda.gov/briefing/organic/*. The briefing room describes characteristics of the U.S. organic farm sector, including estimates of certified organic farmland acreage and livestock, by commodity and by State. The briefing room also features industry data depicting industry growth and sales. Other highlights include ERS publications on organic agriculture and current organic-related activities of ERS researchers.

ERS developed a new set of statistical indicators several years ago—certified organic acreage and livestock estimates—to track the organic farm sector as regulatory and market conditions change. These estimates are based on data from State and private certifiers in the United States and Canada who certify U.S. organic producers. Estimates are posted in the organic briefing room.

Another current research project focuses on risk management in the U.S. organic farm sector. The study will include a review of the economics of organic farming and an analysis of public and private data sources for organic prices and other economic statistics. Focus group discussions were conducted with various types of organic farmers, in different regions of the country, in order to better understand the risk management strategies and needs in the organic farm sector. Previous ERS organic research projects have examined U.S. consumer demand for organic milk, frozen vegetables, and baby food, and the farming practices and socioeconomic characteristics of certified organic fruit and vegetable producers in the United States.

USDA's Foreign Agricultural Service and Organic Agriculture

Contributed by Pam McKenzie and Kelly Stzrelecki, USDA-FAS

The Foreign Agricultural Service (FAS) assists the organic industry with U.S. export programs and services. The FAS International Trade Policy group, in conjunction with the Agricultural Marketing Service, has developed protocols for working with foreign nations to keep organic trade moving as more countries develop organic standards. This program has also worked on many trade issues facing the organic industry including labeling, certification, and market access.

Since 1999, the FAS Commodity and Marketing Program (CMP) area has made over $180,000 of funding available to the Organic Trade Association (OTA) through the Market Access Program (MAP) to conduct marketing efforts in Canada, Europe, and Japan. This support has enabled OTA to develop an extensive export directory titled "Organic Trade Association's Organic Export

Directory." This publication will be available via the Internet and printed in four languages this year. MAP support has also assisted the OTA in conducting market research and using internationally based trade shows to exhibit product and information on U.S. organic products.

CMP's AgExport Services division has published the *Organic Perspectives Newsletter* since 1998 (see *http://www.fas.usda.gov/agx/organics/organics.html*). This bimonthly newsletter contains reports on organics from around the world using U.S. attaché reports, trip reports made by AgExport Services staff, and other sources. The newsletter also covers items of interest concerning the U.S. national organic program and the domestic organic industry. A list of upcoming conferences, trade shows, and other events is included in every issue. In 2002, CMP's AgExport Services Division will be helping OTA establish improved market presence, and FAS hopes to include OTA in some of its other programs soon. Additional programs include the Emerging Markets Program (EMP), Quality Samples Program (QSP), Cochran Program, and Section 108 funding. All programs provide funding to encourage international market development.

USDA's National Agricultural Statistics Service and Organic Agriculture

Contributed by Doug Kleweno USDA-NASS

The National Agricultural Statistics Service (NASS) (see *http://www.usda.gov/nass/*) conducts hundreds of surveys each year and prepares reports that cover virtually every facet of U.S. agriculture—production and supplies of food and fiber, prices paid and received by farmers, farm labor and wages, and other aspects of the industry. In addition, NASS's 45 State Statistical Offices (see *http://www.usda.gov/nass/sso-rpts.htm*) publish data about many of the same topics for local audiences.

Every 5 years, Congress requires NASS to conduct a Census of Agriculture. The Census of Agriculture is the most comprehensive source of data portraying U.S. agriculture. It is the only source of uniform data on agricultural production and operator characteristics for each county, State, and the United States. In the 2002 Census of Agriculture, questions on certified organic production commodities will be included for the first time, along with other new questions on computer/Internet use, production contracts/landlord shares, aquatic plants, acres treated with manure, grain storage capacity, new commodities (bison, deer, elk, llama, emus, and ostriches), and more detailed farm-related income. Report forms for the 2002 Census of Agriculture will be mailed to farm and ranch operators in late December 2002, and NASS will begin release of data starting in spring 2004 in both electronic and print media (see *www.usda.gov/nass/* and click "Census of Agriculture").

USDA's Natural Resources Conservation Service and Organic Agriculture

Contributed by Peter Smith, USDA-NRCS

The Natural Resources Conservation Service (NRCS) (see *http://www.nrcs.usda.gov/*) provides leadership in a partnership effort to help people conserve, maintain, and improve our natural resources and environment. NRCS technical experts help land managers and communities take a comprehensive approach in planning the use and protection of soil, water and related resources on private and non-Federal lands. NRCS assistance is provided through conservation districts that are units of local government created by State law. NRCS works in partnerships with State conservation agencies and other State and local agencies such as resource conservation and development councils and local farmer committees, Federal agencies, tribal governments, and private sector organizations.

NRCS's technical and financial assistance resources are available to all producers, including organic. Almost all of NRCS's technical information, such as soil surveys and standards and specifications for erosion control, wetlands rehabilitation, and wildlife habitat improvement apply to organic and conventional farming.

The Farm Security and Rural Investment Act of 2002 increased the funding for the Environmental Quality Incentives Program, which is intended to improve conservation and sustainability on organic farming operations. In addition, funding for the Farmland Protection Program was greatly increased. This could be significant in protecting organic farming operations from development, especially those close to urban areas. The newly authorized Conservation Security Program can benefit producers who adopt conservation practices.

To highlight its assistance to organic producers, the agency entered into a memorandum of understanding with the Organic Trade Association (OTA). The memorandum establishes a framework for cooperation between NRCS and OTA on program activities that involve the conservation of natural resources specifically related to organic farming. Both the NRCS and the OTA agree to cooperate in developing and implementing farm plans for organic crop production, to encourage the use of demonstrations and field days with organic operations to showcase conservation and organic production, and to share training opportunities, conferences, and newsletters.

Main points covered in the memorandum of understanding between NRCS and the OTA:

NRCS agrees to:

- Provide technical assistance at the national level to develop guidance on natural resources conservation,
- Provide contact information at the State level for councils participating with OTA,
- Furnish conservation technical assistance for conservation measures in organic farm plans,
- Provide soil survey information, soil interpretations, and conservation planning assistance to develop and implement conservation plans consistent with the grower's objectives to achieve sound land use and conservation treatment while producing quality organic crops.

OTA agrees to:

- Discuss conservation technical assistance needs of organic growers annually with NRCS and recommend priorities,
- Recommend conservation planning to growers to aid them in meeting the National Organic Program's crop production standards,
- Inform OTA members of the opportunities and advantages of developing a conservation plan and including it in their organic farm plan.

NRCS assembled and published a 950-page resource manual, "Alternative Enterprise and Agritourism: Farming for Profit and Sustainability," to assist its field staff and partners in helping farmers sustain their operations. A number of opportunities for producing and marketing organic products for direct sale to consumers and for sale to restaurants, wholesalers, and institutions are described. Other technical assistance products include a summarized version of the manual, a CD version of the entire manual, 21 success stories and four information sheets. These activities are coordinated with the NRCS outreach and small farms program activities.

The Hartman Group. *The Organic Consumer Profile*. Bellevue,WA. January 2000.

Klonsky, Karen, Laura Tourte, and David Chaney. "Production Practices and Sample Costs for Organic Processing Tomatoes in the Sacramento Valley" University of California Cooperative Extension Service. Davis,CA. 1993-94a.

Klonsky, Karen, Laura Tourte, and Chuck Ingels. "Production Practices and Sample Costs for Fresh Market Organic Apples, Central Coast," University of California Cooperative Extension Service. Davis, CA. 1993-94b.

Klonsky, Karen, Laura Tourte, and Chuck Ingels. "Production Practices and Sample Costs for Fresh Market Organic Apples, 1993-94, North Coast," University of California Cooperative Extension Service. Davis, CA. 1993-94c.

Klonsky, Karen, and Peter Livingston, 1994. " Alternative Systems Aim to Reduce Inputs, Maintain Profits," *California Agriculture*,Vol. 48, No. 5.

Knoblauch,Wayne A., Rebecca Brown, and Martin Braster. "Organic Field Crop Production:A Review of the Economic Literature,"A.E. Res. 90-10, Department of Agricultural Economics, Cornell University, Ithaca, NY, July 1990.

Loureiro, Maria L. and Susan E. Hine. "Discovering Niche Markets:A Comparison of Willingness to Pay for a Local (Colorado Grown), Organic, and GMO-Free Product." Selected Paper presented at Annual Meeting of the American Agricultural Economics Association, Chicago, IL. 2001.

Loureiro, Maria L., Jill J. McCluskey, and Ron C. Mittlehammer. "Assessing Consumer Preferences for Organic, Eco-labeled, and Regular Apples." *Journal of Agricultural and Resource Economics*. December 2001.

Myers, Steve, and Somlynn Rorie. "Facts and Stats: The Year in Review." *Organic and Natural News*. December 2000.

Natural Foods Merchandiser, various issues.

Nutrition Business Journal, *www.nutritionbusiness.com*

Organic Farming Research Foundation. *Biennial National Organic Farmers'Survey*. 1993, 1995, and 1997.

Organic and Natural News, various issues.

Organic Trade Association. *Manufacturers'Market Survey 2001*. Greenfield, MA. Organic Trade Association. *Manufacturers'Market Survey 1998*. Greenfield, MA. Packaged Facts. *The U.S. Organic Food Market. 2000*. New York, NY.

The Packer. *Fresh Trends: Profile of the Fresh Produce Consumer*. 2002, 2000, 1998, 1996.

Reganold, J.P., J.D. Glover, P.K. Anrews, H.R. Hinman, 2001. "Sustainability of three apple production systems, *Nature*, 410: 926-930.

USDA's Risk Management Agency and Organic Agriculture

Contributed by Sharon Hestvik, USDA-RMA

The Risk Management Agency (RMA) administers the Federal Crop Insurance Corporation (FCIC). RMA offers Federal crop insurance products through a network of private insurance company partners, oversees the creation of new products, seeks enhancements in existing products, ensures the integrity of crop insurance programs, offers outreach programs aimed at underserved communities, and provides risk management education and information (see *www.rma.usda.gov*).

Crop insurance is currently available for 116 different crops in a variety of risk management products to help producers manage the risks in their farming operations. New initiatives resulting from passage of the Agricultural Risk Protection Act of 2000 (ARPA) further contribute to producers'ability to protect their financial stability and comprise a major component of the safety net for agricultural producers. ARPA provisions placed a major emphasis on contracting and partnering for the purpose of developing new risk management tools and also provided for recognition of scientifically sound sustainable and organic farming practices as good farming practices.

Since 2001, RMA has provided coverage for organic farming practices as good farming practices by written agreement. All crops currently covered by RMA are eligible for coverage. Organic farmers are eligible for production losses from damage due to insects, disease, and/or weeds. Coverage is available for both transitional and certified organic acreage, in accordance with approved underwriting guidelines and procedures.

Organic farmers have signed up for written agreements in 19 States on the following crops: apples, almonds, barley, corn, cotton, cranberries, dry beans, flax, grapes, oats, pears, popcorn, sunflowers, soybeans, and wheat.

USDA's Economic Research Service is currently conducting research with RMA that will provide RMA with guidance on developing crop insurance and other risk management tools for scientifically sound sustainable and organic farming practices. Also, a unique new survey, part of a research partnership between RMA, National Agricultural Statistics Service, and several Land Grant Universities, is underway that will provide the data necessary to develop crop insurance products or programs targeted to specialty crop producers. One important aspect of the survey is the section geared to organic producers and their unique practices. Organic producers will be asked to identify the type and percentage of organic acreage planted in 2001, which will allow for comparisons to be made between conventional and organic producers in the ways they use risk management tools.

USDA Publications Related to Organic Agriculture

Contributed by Mary V.Gold, USDA-NAL

USDA Organic Standards and Labels

Frequently Asked Questions (FAQs). National Organic Program,Agricultural Marketing Service, U.S. Department of Agriculture, 2002. Available at *http://www.ams.usda.gov/nop/nop2000/nop2/FAQ.htm*

Organic Food Standards and Labels: The Facts. National Organic Program,Agricul-tural Marketing Service, U.S. Department of Agriculture, 2002. Available at *http://www.ams.usda.gov/nop/consumerbrochure.htm*

Agronomic Research Related to Organic Agriculture

"Research for MeBr Alternatives May Benefit Organic Growers."Agricultural Research Service, U.S. Department of Agriculture, Methyl Bromide Alternatives, 5, no. 4, October 1999. Available at *http://www.ars.usda.gov/is/np/mba/oct99/organic.htm*

Comis, Don. "Soil Fungi Critical to Organic Success."Agricultural Research Service, U.S. Department of Agriculture,Agricultural Research, May 2001. Available at *http://www.ars.usda.gov/is/AR/archive/may01/fungi0501.htm*
Hardin, Ben. "An Organic Cure for a Turfgrass Disease?"Agricultural Research Service, U.S. Department of Agriculture,Agricultural Research,August 2001. Available at *http://www.ars.usda.gov/is/AR/archive/aug01/turf0801.htm*

Jawson, Michael D. "Forum—Adding OOMPH to Organic Farming."Agricultural Research Service, U.S. Department of Agriculture, *Agricultural Research*, February 2002. Available at *http://www.ars.usda.gov/is/AR/archive/feb02/form0202.htm*

Suszkiw, Jan. "Fungal Protein Slows Broadleaf Weeds."Agricultural Research Service, U.S. Department of Agriculture, *Agricultural Research*, February 2000. Available at *http://www.ars.usda.gov/is/AR/archive/feb00/weeds0200.htm*

Wood, Marcia, and others. "Organic Grows on America!"Agricultural Research Service, U.S. Department of Agriculture, *Agricultural Research*, February 2002. Available at *http://www.ars.usda.gov/is/AR/archive/feb02/amer0202.htm*

Organic Information Resources

Gold, Mary V. *Organic Foods: Markets and Marketing—Recent Books, Articles, News Items, Reports, and Videos.* National Agricultural Library,Agricultural Research Service, U.S. Department of Agriculture, 2001. Available at *http://www.nal.usda.gov/afsic/AFSIC_pubs/orgmrkt.htm*

Gold, Mary V. *Organic Information Resources:What Are They? Where Are They? How Can I Find Them?* National Agricultural Library,Agricultural Research Service, U.S. Department of Agriculture, 2001. Available at *http://www.nal.usda.gov/afsic/ofp/orgfind.htm*

Gold, Mary V. *Organic Production: Economic Aspects. Books, Articles, and Videocassettes, 1991 - March 1997.* National Agricultural Library,Agricultural Research Service, U.S. Department of Agriculture, 1997. Available at *http://www.nal.usda.gov/afsic/AFSIC_pubs/organic1.htm*

Gold, Mary V. *Organic Production: Recent Publications and Current Information Sources.* (Special Reference Briefs, no. SRB 96-07) National Agricultural Library, Agricultural Research Service, U.S. Department of Agriculture, 1996. Available at *http://www.nal.usda.gov/afsic/AFSIC_pubs/srb96-07.htm*

Gold, Mary V. *Organically Produced Foods: Nutritive Content.* (Special Reference Briefs, no. SRB 2000-03) National Agricultural Library,Agricultural Research Service, U.S. Department of Agriculture, 2000. Available at *http://www.nal.usda.gov/afsic/AFSIC_pubs/srb0003.htm*

Organic Production Bulletins and Technical Information

Explore Organic Production. (Farming for Profit, Stewardship & Community Tip Sheet, no. 4) Sustainable Agriculture Network, Sustainable Agriculture Research and Education, Cooperative State Research, Education, and Extension Service, U.S. Department of Agriculture, 1999. Available at *http://www.sare.org / tipsheet/tip4.htm*

Naturalize Your Farming System: A Whole-Farm Approach to Managing Pests. (Information Bulletin) Sustainable Agriculture Network, Sustainable Agriculture

Research and Education, Cooperative State Research, Education, and Extension Service, U.S. Department of Agriculture, 2000.

Profitable Poultry: Raising Birds on Pasture. (Information Bulletin) Sustainable Agriculture Network, Sustainable Agriculture Research and Education, Cooperative State Research, Education, and Extension Service, U.S. Department of Agriculture, 2002. Available at *http://www.sare.org/bulletin/poultry/*

Smith, Miranda and Members of the Northeast Organic Farming Association and Cooperative Extension. *The Real Dirt: Farmers Tell about Organic and Low-Input Practices in the Northeast.* Northeast Region, Sustainable Agriculture Research and Education Program, Cooperative State Research, Education, and Extension Service, U.S. Department of Agriculture, 1994.

Economic Research and Analysis

Anton, Julie, and B. Frazao. "Organic Certification: Standards in the Works." Economic Research Service, U.S. Department of Agriculture, *Agricultural Outlook*, AO-199 1993: 26-31.

Dimitri, Carolyn, and Nessa Richman. "Organic Foods: Niche Marketers Venture into the Mainstream." Economic Research Service, U.S. Department of Agriculture, *Agricultural Outlook,* June 2000-July 2000. Available at *http://www.ers.usda.gov/publications/AgOutlook/Jun2000/ao272f.pdf*

Dunn, Julie. "Organic Foods Find Opportunity in the Natural Food Industry." Economic Research Service, U.S. Department of Agriculture, *FoodReview*, 18, no. 3, 1995: 7-12.

Fernandez-Cornejo, Jorge, Renata Penn, and Doris Newton. "Organic Fruit Growers Survey." Economic Research Service, U.S. Department of Agriculture, *Updates on Agricultural Resources and Environmental Indicators, June 1997.* Available at *http://www.ers.usda.gov/publications/arei/97upd/upd97-4.pdf*

Fernandez-Cornejo, Jorge, Renata Penn, and Doris Newton. "Organic Vegetable Growers Surveyed in 1994." Economic Research Service, U.S. Department of Agriculture, *Updates on Agricultural Resources and Environmental Indicators,* May 1996. Available at *http://www.ers.usda.gov/publications/arei/96upd/upd96-4.pdf*

Glaser, Lewrene, and others. "Demand for Frozen Vegetables:A Comparison of Organic and Conventional Products." Economic Research Service, U.S.

Department of Agriculture, *Vegetables and Specialties,* November 1998. Available at *http://usda.mannlib.cornell.edu/reports/erssor/specialty/vgs-bb/1998/vgs276f.asc*

Golan, Elise, and others. *Economics of Food Labeling.* (Agricultural Economic Report No. 793) Economic Research Service, U.S. Department of Agriculture, 2001. Available at *http://www.ers.usda.gov/publications/aer793/*

Greene, Catherine and Thomas Dobbs. "Organic Wheat Production in the United States: Expanding Markets and Supplies." Economic Research Service, U.S. Department of Agriculture, *Wheat Situation and Outlook Yearbook,* March 2001. Available at *http://www.ers.usda.gov/publications/so/view.asp?f=field/whs-bby/whs2001.pdf*

Greene, Catherine. "U.S. Organic Agriculture Gaining Ground." Economic Research Service, U.S. Department of Agriculture, *Agricultural Outlook,* AGO-270 April 2000. Available at *http://www.ers.usda.gov/publications/agoutlook/apr2000/ao270d.pdf*

Greene, Catherine. *U.S. Organic Farming Emerges in the 1990s: Adoption of Certified Systems.* (Agriculture Information Bulletin No. 770) Economic Research Service, U.S. Department of Agriculture, 2001. Available at *http://www.ers.usda.gov/publications/aib770/*

Greene, Catherine, Carolyn Dimitri, and Nessa Richman. "Organic Marketing Features Fresh Food and Direct Exchange." Economic Research Service, U.S. Department of Agriculture, *FoodReview*, 24, no. 1, January 2001-April 2001. Available at *http://www.ers.usda.gov/publications/FoodReview/jan2001/frv24I1f.pdf*

Greene, Catherine, and Linda Calvin. "'Organically Grown'Vegetables: U.S. Acreage and Markets Expand During the 1990s." Economic Research Service, U.S. Department of Agriculture, *Vegetables and Specialties Situation and Outlook*, May 1997. Available at *http://usda.mannlib.cornell.edu/reports/erssor/specialty/vgs-bb/1997/ vgs271f.asc*

Harris, J. Michael. "Consumers Pay a Premium for Organic Baby Foods." Economic Research Service, U.S. Department of Agriculture, *FoodReview*, October 1997. Available at *http://www.ers.usda.gov/publications/foodreview/aug1997/may97d.pdf*

Sok, Emy, and Lewrene Glaser. "Tracking Wholesale Prices for Organic Produce." Economic Research Service, U.S. Department of Agriculture, *Agricultural Outlook,* AGO-285 October 2001. Available at *http://www.ers.usda.gov/publications/AgOutlook/oct2001/*

International Markets

Achilles, Dietmar. "Germany: The Customer-Competitor Duality." Foreign Agricultural Service, U.S. Department of Agriculture, *AgExporter*, June 2000. Available at *http://www.fas.usda.gov/info/agexporter/2000/june/germany.htm*
Attaché Reports. Foreign Agricultural Service, U.S. Department of Agriculture, 1995- present. Available at *http://www.fas.usda.gov/agx/organics/attache.htm*

Good, Dale L. "Singapore: A Growing Market for U.S. Health Foods." Foreign Agricultural Service, U.S. Department of Agriculture, *AgExporter*, September 1998. Available at *http://www.fas.usda.gov/info/agexporter/1998/singapor.html*

Jones, Jennifer. "The U.K. Organic Sector: Go for It!" Foreign Agricultural Service, U.S. Department of Agriculture, *AgExporter*, June 2000. Available at *http://www.fas.usda.gov/info/agexporter/2000/june/uk.htm*

Lee, Jill. "Five Tips for Organic Trade Success." Foreign Agricultural Service, U.S. Department of Agriculture, *AgExporter*, June 2000. Available at *http://www.fas.usda.gov/info/agexporter/2000/june/fivetips.htm*

Lee, Jill. "Organic and Hispanic, These West Coast Chipsters Are Bagging Exports." Foreign Agricultural Service, U.S. Department of Agriculture, *AgExporter*, December 1999. Available at *http://www.fas.usda.gov/info/agexporter/1999/organic.html*

Lee, Jill. "Wizard's Cauldron Conjures International Markets." Foreign Agricultural Service, U.S. Department of Agriculture, *AgExporter*, June 2000. Available at *http://www.fas.usda.gov/info/agexporter/2000/june/wizards.htm*

Lin, Henry. "Organic Food and Vegetable Products, a Sino- the Times." Foreign Agricultural Service, U.S. Department of Agriculture, *AgExporter*, March 1998. Available at *http://www.fas.usda.gov/info/agexporter/1998/organic.html*

Myles, George C. "Defining Moment: Canadian Organic Regs Are Coming." Foreign Agricultural Service, U.S. Department of Agriculture, *AgExporter*, November 1997. Available at *http://www.fas.usda.gov/info/agexporter/1997/defining.html*

Organic Perspectives Newsletter. Foreign Agricultural Service, U.S. Department of Agriculture, 1998- present. Available at *http://www.fas.usda.gov/agx/organics/organics. html*

Perng, Chiou Mey. "Organic Food Trend Wins in Taiwan." Foreign Agricultural Service, U.S. Department of Agriculture, *AgExporter*, September 1998. Available at *http://www.fas.usda.gov/info/agexporter/1998/orgtaiw.html*

Perng, Chiou Mey. "Organic Products Finding a Ready Market on Taiwan." Foreign Agricultural Service, U.S. Department of Agriculture, *AgExporter,* June 2000. Available at *http://www.fas.usda.gov/info/agexporter/2000/june/organic3.htm*

Piason, Frank J. "France: More Than a Nice Niche?" Foreign Agricultural Service, U.S. Department of Agriculture, *AgExporter,* June 2000. Available at *http://www.fas.usda.gov/info/agexporter/2000/june/france.htm*

Schmaedick, Melissa. "Danish Organic Dairy Production." Foreign Agricultural Service, U.S. Department of Agriculture, *FAS Division Reports*, July 2000. Available at *http://www.fas.usda.gov/dlp/circular/2000/00-07Dairy/danish.html*

Sharpless, Mattie R., and Roselyne Gauthier. "France's Baby Food Boom." Foreign Agricultural Service, U.S. Department of Agriculture, *AgExporter*, February 1998. Available at *http://www.fas.usda.gov/info/agexporter/1998/frances.html*

Thorburn II, W. Garth. "Organic Food Trend Sprouts in Hong Kong." Foreign Agricultural Service, U.S. Department of Agriculture, *AgExporter*, September 1998. Available at *http://www.fas.usda.gov/info/agexporter/1998/orgHoKoc.html*

Foreign Agricultural Service. "Practical Contacts for Organic Producers and Exporters." U.S. Department of Agriculture, *AgExporter,* June 2000. Available at *http://www.fas.usda.gov/info/agexporter/2000/june/practica.htm*

Foreign Agricultural Service. "Timeline of Organic Milestones." U.S. Department of Agriculture, *AgExporter*, June 2000. Available at *http://www.fas.usda.gov/info/agexporter/2000/june/timeline.htm*

van der Harst-Collaris, Tatiana, and Laura Gabel Scandurra. "Dutch Organic Food Market Offers All-Natural Potential for U.S. Firms." U.S. Department of

Agriculture, *AgExporter*, August 1997. Available at *http://www.fas.usda.gov/info/agexporter/1997/dutch.html*

Yuen, Carolyn. "Lacking a Domestic Source, Hong Kong Must Import Organics." Foreign Agricultural Service, U.S. Department of Agriculture, *AgExporter*, June 2000. Available at *http://www.fas.usda.gov/info/agexporter/2000/june/lackinga.htm*

Zygmont, Janise. "Organic Markets Offer U.S. Agriculture Current and Future Sales Opportunities." Foreign Agricultural Service, U.S. Department of Agriculture, *AgExporter*, June 2000. Available at *http://www.fas.usda.gov/info/agexporter/2000/june/organic.htm*

Zygmont, Janise. "Organic Trade Association Sees Opportunities in Exports." Foreign Agricultural Service, U.S. Department of Agriculture, *AgExporter*, June 2000. Available at *http://www.fas.usda.gov/info/agexporter/2000/june/ organic2 .htm*

Zygmont, Janise. "U.S. Organic Fruit: Export Opportunities and Competition in the International Market." Foreign Agricultural Service, U.S. Department of Agriculture, *FAS Division Reports*, 2001. Available at *http://www.fas.usda.gov /agx/organics/speech.htm*

BIBLIOGRAPHY

Assadian, N., L. Corral Esparza, and C. Ponce, 1999. "Organic Cotton Production -1993 to 1997," Draft research bulletin, Texas A&M Research Farm, El Paso,TX

Batte, Marvin T., D. Lynn Forster, and Fred J. Hitzhusen, 1993. "Organic Agriculture in Ohio: An Economic Perspective," *Journal of Production Agriculture*,Vol. 6, No. 4.

Bertramsen, Sherry K., and Thomas L. Dobbs. "Comparison of Prices for 'Organic' and 'Conventional'Grains and Soybeans in the Northern Great Plains and Upper Midwest: 1995 to 2000. Econ Pamphlet 2001-1. South Dakota State University.

Bertramsen, Sherry K., and Thomas L. Dobbs. "An Update on Prices of Organic Crops in Comparison to Conventional Crops." *Economics Commentator.* February 22, 2002. South Dakota State University.

Born, Holly, and Preston Sullivan. *Marketing Organic Grains. ATTRA Publication*. February 2002. Arkansas.

Bunin, Lisa. "Center Grants Support Student Fieldwork," *The Cultivar: The Center for Agroecology & Sustainable Food Systems,* University of California, Santa Cruz,Vol. 28, No. 2, Fall/Winter. 2000.

Dimitri, Carolyn, and Nessa J. Richman. *Organic Food Markets in Transition.* Policy Studies Report No. 14. Henry A. Wallace Center for Agricultural and Environmental Policy,Winrock International. Greenbelt, MD. April 2000.

Dobbs, Thomas L., Dillon M. Feuz, and Richard C. Shane. "Expanding the Organic Food and Agriculture System in the U.S.'s Upper Midwest: Strategies and Lessons of a Pilot Project." Report for The Pew Charitable Trusts. Brookings, South Dakota. August 1999.

Estes, Edmund, and V. Kerry Smith. "Price, Quality, and Pesticide-Related Health Risk Considerations in Fruit and Vegetable Purchases:An Hedonic Analysis of Tucson, Arizona Supermarkets," *Journal of Food Distribution Research* 27(3): 8-17, Fall 1996.

Fernandez-Cornejo, Jorge, Catherine Greene, Renata Penn, and Doris Newton. "Organic Vegetable Production in the U.S.: Certified Growers and their Practices," *American Journal of Alternative Agriculture*,Vol. 13, No. 2, 1998.

Food Marketing Institute. *Trends in the United States: Consumer Attitudes & the Supermarket, 2001.*

Gegner, Lance. *Considerations in Organic Hog Production.* ATTRA's Organic Matters Series. National Center for Appropriate Technology. Fayetteville,Arkansas. July 2001.

Glaser, Lewrene, and Gary D. Thompson, "Demand for Organic and Conventional Frozen Vegetables." Selected Paper presented at Annual Meeting of the American Agricultural Economics Association, Nashville, TN. August, 1999.

Glaser, Lewrene, and Gary D. Thompson, "Demand for Organic and Conventional Beverage Milk " Selected Paper presented at Annual Meeting of the Western Agricultural Economics Association,Vancouver, British Columbia. June 29-July 1, 2000.

Govidnasamy, Ramu, and John Italia. "Predicting Willingness to Pay a Premium for Organically Grown Fresh Produce." *Journal of Food Distribution Research.* 1990. Volume 30, Number 2, pages 44-53.

Granato, Heather. "Mapping the Changing Organic World." *Organic and Natural News.* September/October 2001.

Greene, Catherine. *U.S. Organic Farming Emerges in the 1990s: Adoption of Certified Systems.* Economic Research Service, U.S. Department of Agriculture. Agriculture Information Bulletin No. 770, June 2001.

Reicks, Marla, Patricia Splett, and Amy Fishman. "Shelf Labeling of Organic Foods: Customer Response in Minnesota Grocery Stores." *Journal of Food Distribution Research.* 1990. Volume 30, Number 2, pages 11-23.

Sok, Emy, and Lewrene Glaser. "Tracking Wholesale Prices for Organic Produce." U.S. Department of Agriculture, Economic Research Service, *Agricultural Outlook*, October 2001.

Swezey, Sean, Jim Rider, Matthew Werner, Marc Buchanan, Jan Allison, and

Stephen Gliessman, 1994. "Granny Smith conversions to organic show early success," *California Agriculture,* Vol. 48.

Thompson, Gary, and Lewrene Glaser. "National Demand for Organic and Conventional Baby Foods," Selected Paper presented at Annual Meeting of the Western Agricultural Economics Association, Logan, Utah, July 8-11, 2001.

Thompson, Gary, and Julia Kidwell. "Explaining the Choice of Organic Produce: Cosmetic Defects, Prices, and Consumer Preferences." *American Journal of Agricultural Economics* v80, n2 (May 1998): 277-8

United States Department of Agriculture,Agricultural Marketing Service. "National Organic Program, Final Rule," *Federal Register* 7CFR Pt 205, December 21, 2000.

United States Department of Agriculture, National Agricultural Library,Alternative Farming Systems Information Center, "Community Supported Agriculture." See *http://www.nal.usda.gov/afsic/.*

Walnut Acres survey. *www.walnutacres.com.* 2001.

Welsh, Rick. *The Economics of Organic Grain and Soybean Production in the Midwestern United States.* Policy Studies Report No. 13, Henry A. Wallace Institute for Alternative Agriculture. Greenbelt, MD. May 1999.

ACKNOWLEDGMENTS

The authors are grateful for contributions from several people: Tom Bewick, Jerry DeWitt, Mary Gold, Sharon Hestvik, Mike Jawson, Doug Kleweno, Pam McKenzie, Kelly Strzelecki, Peter Smith, and Demaris Wilson. We extend a special thanks to Mary Gold for providing a list of articles on organic agriculture published by USDA. The authors gratefully acknowledge the helpful review comments of Mary Bohman,William Chambers, Robert Dismukes, Tom Dobbs, Kitty Smith, Gary Thompson, Utpal Vasavada, Rick Welsh, and Keith Wiebe. We are also indebted to Susan DeGeorge for the graphic design work she completed for this report and to Tom McDonald for valuable editorial assistance.

REFERENCES

[1] National Organic Program, Final Rule, *Federal Register* 7CFR Pt 205, December 21, 2000.

[2] Traditionally, USDA has regulated labeling of all meat products and FDA has regulated labeling of nonmeat animal products and other food products. The new USDA organic label, however, will apply to all food products.

[3] Consumers gave more than one reason for purchasing organic food. Thus, the percentages sum to more than 100.

[4] Wild crops are plants or portions of plants that are harvested from land that is not maintained under cultivation or other agricultural management.

In: Organic Agriculture in the U.S.
Editor: Alison J. Wellson, pp. 191-222
ISBN: 978-1-60021-305-2

Chapter 6

What Does "Certified Organic" Mean?

Gary Thompson

Q. What does "certified organic" mean?

A. "Certified organic" means that agricultural products have been grown and processed according to USDA's national organic standards and certified by USDA-accredited State and private certification organizations. Certifying agents review applications from farmers and processors for certification eligibility, and qualified inspectors conduct annual onsite inspections of their operations. Inspectors talk with operators and observe their production and processing practices to determine if they are in compliance with organic standards that, for example, virtually prohibit synthetic pesticide use in crop production and require outdoor access for animals in livestock production.

Private organizations, mostly nonprofits, began developing certification standards in the early 1970s as a way to support organic farming and thwart consumer fraud, and some States began offering organic certification services in the late 1980s for similar reasons. Over 50 organic certification organizations, including 14 State programs, conducted third-party certification in 2001.

Congress passed the Organic Foods Production Act of 1990 to establish national standards for organically produced commodities. This legislation requires that all except the smallest organic growers will have to be certified by a State or private agency accredited under the national standards that are administered by USDA's National Organic Program. The State and private groups that currently certify growers are expected to seek accreditation by USDA when the national program is implemented.

Q. What's new in USDA's organic programs?

A. On October 21, 2002, USDA's regulations on production, handling, and processing of organically grown agricultural products came into effect. The entire Final Rule and summary fact sheets are available from USDA's National Organic Program.

The 2002 Farm Act includes several first-time research and technical assistance provisions to assist organic crop and livestock producers with production and marketing. The Act authorizes $15 million in new funding for organic production system research and $15 million for a national cost-share program to help defray the costs of certification incurred by organic producers.

Previously, USDA announced on May 24, 2001, that an organic certification cost-share program with $1 million in fiscal year 2001 funds was available to share the cost of organic certification in the states of Connecticut, Delaware, Maine, Maryland, Massachusetts, Nevada, New Hampshire, New Jersey, New York, Pennsylvania, Rhode Island, Utah, Vermont, West Virginia, and Wyoming. Each producer is eligible for reimbursement of up to 70 percent of its cost of certification, not to exceed $500.

On January 17, 2001, USDA's Risk Management Agency released Bulletin No. MGR-01-004 implementing the organic farming practices section of the Agricultural Risk Protection Act of 2000. Section 123 requires that organic farming practices be recognized as good farming practices.

Q. How much acreage in the United States is certified organic?

A. Farmers in 48 States used organic production systems and third-party organic certification services on 2.3 million acres of farmland in 2001 and were raising certified organic livestock in nearly three-quarters of the States, according to recent ERS data on organic production. Over half of the farmland was used for growing crops, with California, North Dakota, Minnesota, Wisconsin, Iowa, and Montana as the top producers. Colorado and Texas had the most organic pasture and rangeland.

Organic farming has made deeper inroads in the fruit, vegetable, and specialty grain sectors than in other farm sectors. While only 0.2 percent of U.S. corn and 0.3 percent of soybeans were grown under certified organic farming systems in 2001, about 1 percent of oats, dry beans, tomatoes, grapes, and citrus were grown organically. About 2 percent of dry peas and lentils, 3 percent of apples, 4 percent of carrots, and 5 percent of lettuce was organic.

Q. Where can I receive technical assistance on organic production, certification, and marketing?

A. Staff at the Appropriate Technology Transfer for Rural Areas (ATTRA)—a USDA-funded, nonprofit information center on sustainable farming—provide technical assistance, publications, and resources on organic agriculture and other topics to farmers, Extension agents, market gardeners, agricultural researchers, and others across the United States. Staff may be reached toll-free at 1-800-346-9140.

The Organic Farming Research Foundation is a California-based nonprofit that provides funding for farmer-scientist research collaborations and conducts a comprehensive survey of certified organic farmers.

USDA's Sustainable Agriculture Research and Education Program (SARE) provides program contacts, publications, funded projects, funding opportunities, and an interactive calendar of sustainable agriculture events. Regional organic farming conferences are hosted by a variety of industry, government, and academic sponsors and may be a useful source of information. USDA's Alternative Farming Systems Information Center provides a wide range of information resources on organic farming and sustainable agriculture.

Some State cooperative extension services—Nebraska Cooperative Extension, for example—provide information on organic transition and certification. A list of State organic certifiers and State contacts is available from USDA's National Organic Program.

Q. What is the size of the U.S. market for organic foods?

A. USDA does not have official statistics on U.S. organic retail sales, but information is available from industry sources. These sources sometimes produce conflicting statistics. New Hope Natural Media's Natural Foods Merchandiser (NFM), an industry trade publication, published estimates of U.S. organic sales through all channels (including exports) from 1990 through 1996, which showed sales growing 20-25 percent annually and reaching \$3.3 billion in 1996. New Hope Natural Media subsequently revised its survey methods, and reported statistics about the industry in the Nutrition Business Journal (NBJ). The NBJ reported industry sales of \$3.47 billion in 1997, \$4.15 billion in 1998, \$4.89 billion in 1999, \$5.9 billion in 2000, and \$7.10

billion in 2001. Another market research firm, Packaged Facts, reported total organic retail sales through all outlets between 1996 and 2000. The firm estimates that organic food sales were $7.8 billion in 2000, a 20-percent increase over 1999 sales.

Natural food stores began growing in size and product selection during the early 1990's, making organic food increasingly available to consumers. Conventional supermarkets and mainstream stores began integrating a wider selection of organic products in the late 1990's and now account for 49 percent of total retail sales, about the same as natural food stores (48 percent), according to Packaged Facts. The firm also estimates that 3 percent of U.S. organic food sales occur through farmers markets and consumer subscription arrangements. NFM estimated that exports accounted for about 5 percent of total U.S. organic food sales in the early 1990's, and a recent USDA-funded study indicated that exports are still about 5 percent of total U.S. sales.

Fresh fruits and vegetables have been the top selling category of organically grown food since the organic food industry started retailing products over three decades ago, and they are still outselling other food categories, according to the Nutrition Business Journal. Produce accounted for 43 percent of U.S. organic food sales in 2002, followed by bread and grains (13 percent), packaged and prepared foods (11 percent), dairy (11 percent), beverages (11 percent), soy products (7 percent), snack foods (3 percent) and meat, fish, and poultry (3 percent).

One of the most striking differences between conventional and organic food marketing is the use of direct markets, such as farmers' markets, farmstands, roadside stands, farm subscriptions, and mail- order sales. Cornell University recently estimated that only about 1.6 percent of U.S. fresh produce sales occur directly between producers and consumers. For organic sales, however, direct markets accounted for between 17 and 22 percent of total organic sales during 1990-96, as tracked by NFM surveys. USDA surveys of organic fruit and organic vegetable growers indicate an even higher percentage using direct marketing as their primary sales outlet.

The number of farmers' markets in the United States has grown steadily from 1,755 markets in 1994, when USDA began to track them, to 2,863 in 2000. Farmers' markets are scattered across the country, but have heavy concentrations in the Northeast, Upper Midwest, and along the West Coast.

Q. Where can I find information on prices for organic commodities?

A. USDA does not systematically report organic prices. USDA' Agricultural Marketing Service Market News Reports publishes daily wholesale prices from a number of terminal markets around the country. A few of the reports contain prices for several organic vegetables, but most do not. The organic prices reported will vary by market and by vegetable. ERS compiled organic prices from two of these markets (Boston and San Francisco) for 1995-2001.

ERS tracked wholesale organic price premiums for two vegetables (romaine lettuce and cherry tomatoes) between 1989 and 1992—based on a price report published by the nonprofit Community Alliance for Family Farmers—and found annual average prices that were generally double conventional prices, with wide variation on a weekly basis.

Several private firms collect and sell organic price data. Contact information for these firms is available through Resources for Organic Marketing (Appropriate Technology Transfer for Rural Areas). A private firm based in Florida began publishing a weekly organic price report in the early 1990's, Organic Food Business News FAX Bulletin (OFBN), containing farmgate prices for grains and produce. In 2003, The Rodale Institute began publishing online weekly reports with terminal market and other wholesale market prices and for selected large-scale retail prices for a wide variety of organic products, including fruits and vegetables, dairy, and grains.

ERS reported that monthly farmgate price premiums for several major fruits and vegetables consistently exceeded 100 percent between 1992 and 1996 based on these data (see Organic Farming and Marketing Research - New Partnerships and Priorities, pp. 47-52), and that supermarket scanner data showed similar results for frozen vegetables. ERS also reported that for organic milk, scanner data showed a 60-percent premium over conventional milk brands from 1997-99. For farmgate grain premiums, Tom Dobbs at South Dakota State University tracked organic corn, soybean, wheat, and oat prices for 1993-99 using OFBN data and found organic farmgate premiums that consistently exceeded 50 percent for these crops.

Q. What types of consumers buy organic foods?

A. Numerous studies have been conducted by researchers in the public and private sectors on the buying habits and demographics of consumers of organic foods. Specific results have varied depending on the type of survey, sample size, and geographic coverage. However, a few general themes have emerged. Consumers prefer organically produced food because of perceived health attributes and concern about pesticide residues, the environment (such as soil and water quality and wildlife habitat), and farmworker safety.

ORGANIC FOODS PRODUCION ACT OF 1990 [AS AMENDED THROUGH PUBLIC LAW 109-97, NOV. 10, 2005]

Title XXI of the Food, Agriculture, Conservation, and Trade Act of 1990 (Public Law 101-624)

TITLE XXI-ORGANIC CERTIFICATION

Sec. 2101. [7 U.S.C. 6501 Note] Short Title

This title may be cited as the "Organic Foods Production Act of 1990".

Sec. 2102. [7 U.S.C. 6501] Purposes

It is the purpose of this title-

(1) to establish national standards governing the marketing of certain agricultural products as organically produced products;
(2) to assure consumers that organically produced products meet a consistent standard; and
(3) to facilitate interstate commerce in fresh and processed food that is organically produced.

Sec. 2103. [7 U.S.C. 6502] Definitions

As used in this title:

(1) AGRICULTURAL PRODUCT.-The term "agricultural product" means any agricultural commodity or product, whether raw or processed, including any commodity or product derived from livestock that is marketed in the United States for human or livestock consumption.

(2) BOTANICAL PESTICIDES.-The term "botanical pesticides" means natural pesticides derived from plants.

(3) CERTIFYING AGENT.-The term "certifying agent" means the chief executive officer of a State or, in the case of a State that provides for the Statewide election of an official to be responsible solely for the administration of the agricultural operations of the State, such official, and any person (including private entities) who is accredited by the Secretary as a certifying agent for the purpose of certifying a farm or handling operation as a certified organic farm or handling operation in accordance with this title.

(4) CERTIFIED ORGANIC FARM.-The term "certified organic farm" means a farm, or portion of a farm, or site where agricultural products or livestock are produced, that is certified by the certifying agent under this title as utilizing a system of organic farming as described by this title.

(5) CERTIFIED ORGANIC HANDLING OPERATION.-The term "certified organic handling operation" means any operation, or portion of any handling operation, that is certified by the certifying agent under this title as utilizing a system of organic handling as described under this title.

(6) CROP YEAR.-The term "crop year" means the normal growing season for a crop as determined by the Secretary.

(7) GOVERNING STATE OFFICIAL.-The term "governing State official" means the chief executive official of a State or, in the case of a State that provides for the Statewide election of an official to be responsible solely for the administration of the agricultural operations of the State, such official, who administers an organic certification program under this title.

(8) HANDLE.-The term "handle" means to sell, process or package agricultural products.

(9) HANDLER.-The term "handler" means any person en-gaged in the business of handling agricultural products, except such term shall not include final retailers of agricultural products that do not process agricultural products.

(10)HANDLING OPERATION.-The term "handling operation" means any operation or portion of an operation (except final retailers of agricultural products that do not process agricultural products) that-

(A) receives or otherwise acquires agricultural products; and
(B) processes, packages, or stores such products.

(11) LIVESTOCK.-The term "livestock" means any cattle, sheep, goats, swine, poultry, equine animals used for food or in the production of food, fish used for food, wild or domesticated game, or other nonplant life.

(12) NATIONAL LIST.-The term "national list" means a list of approved and prohibited substances as provided for in section 2118.

(13) ORGANIC PLAN.-The term "organic plan" means a plan of management of an organic farming or handling operation that has been agreed to by the producer or handler and the certifying agent and that includes written plans concerning all aspects of agricultural production or handling described in this title including crop rotation and other practices as required under this title.

(14) ORGANICALLY PRODUCED.-The term "organically produced" means an agricultural product that is produced and handled in accordance with this title.

(15) PERSON.-The term "person" means an individual, group of individuals, corporation, association, organization, cooperative, or other entity.

(16) PESTICIDE.-The term "pesticide" means any substance which alone, in chemical combination, or in any formulation with one or more substances, is defined as a pesticide in the Federal Insecticide, Fungicide, and Rodenticide Act (7 U.S.C. 136 et seq.).

(17) PROCESSING.-The term "processing" means cooking, baking, heating, drying, mixing, grinding, churning, separating, extracting, cutting, fermenting, eviscerating, pre-serving, dehydrating, freezing, or otherwise manufacturing, and includes the packaging, canning, jarring, or otherwise en-closing food in a container.

(18) PRODUCER.-The term "producer" means a person who engages in the business of growing or producing food or feed.

(19) SECRETARY.-The term "Secretary" means the Secretary of Agriculture.

(20) STATE ORGANIC CERTIFICATION PROGRAM.-The term "State organic certification program" means a program that meets the requirements of section 2107, is approved by the Secretary, and that is designed to ensure that a product that is sold or labeled as "organically produced" under this title is produced and handled using organic methods.

(21) SYNTHETIC.-The term "synthetic" means a substance that is formulated or manufactured by a chemical process or by a process that chemically changes a substance extracted from naturally occurring plant,

animal, or mineral sources, except that such term shall not apply to substances created by naturally occurring biological processes.

Sec. 2104. [7 U.S.C. 6503] National Organic Production Program

(a) IN GENERAL.-The Secretary shall establish an organic certification program for producers and handlers of agricultural products that have been produced using organic methods as provided for in this title.

(b) STATE PROGRAM.-In establishing the program under sub-section (a), the Secretary shall permit each State to implement a State organic certification program for producers and handlers of agricultural products that have been produced using organic methods as provided for in this title.

(c) CONSULTATION.-In developing the program under subsection (a), and the National List under section 2118, the Secretary shall consult with the National Organic Standards Board established under section 2119.

(d) CERTIFICATION.-The Secretary shall implement the program established under subsection (a) through certifying agents. Such certifying agents may certify a farm or handling operation that meets the requirements of this title and the requirements of the organic certification program of the State (if applicable) as an organically certified farm or handling operation.

Sec. 2105. [7 U.S.C. 6504] National Standards for Organic Production

To be sold or labeled as an organically produced agricultural product under this title, an agricultural product shall-

(1) have been produced and handled without the use of synthetic chemicals, except as otherwise provided in this title;

(2) except as otherwise provided in this title and excluding livestock, not be produced on land to which any prohibited sub-stances, including synthetic chemicals, have been applied during the 3 years immediately preceding the harvest of the agricultural products; and

(3) be produced and handled in compliance with an organic plan agreed to by the producer and handler of such product and the certifying agent.

Sec. 2106. [7 U.S.C. 6505] Compliance Requirements

(a) DOMESTIC PRODUCTS.-

(1) IN GENERAL.-On or after October 1, 1993-

(A) a person may sell or label an agricultural product as organically produced only if such product is produced and handled in accordance with this title; and

(B) no person may affix a label to, or provide other market information concerning, an agricultural product if such label or information implies, directly or indirectly, that such product is produced and handled using organic methods, except in accordance with this title.

(2) USDA STANDARDS AND SEAL.-A label affixed, or other market information provided, in accordance with paragraph (1) may indicate that the agricultural product meets Department of Agriculture standards for organic production and may incorporate the Department of Agriculture seal.

(b) IMPORTED PRODUCTS.-Imported agricultural products may be sold or labeled as organically produced if the Secretary determines that such products have been produced and handled under an organic certification program that provides safeguards and guidelines governing the production and handling of such products that are at least equivalent to the requirements of this title.

(c) EXEMPTIONS FOR PROCESSED FOOD.-Subsection (a) shall not apply to agricultural products that-

(1) contain at least 50 percent organically produced ingredients by weight, excluding water and salt, to the extent that the Secretary, in consultation with the National Organic Standards Board and the Secretary of Health and Human Services, has determined to permit the word "organic" to be used on the principal display panel of such products only for the purpose of describing the organically produced ingredients; or

(2) contain less than 50 percent organically produced ingredients by weight, excluding water and salt, to the extent that the Secretary, in consultation with the National Organic Standards Board and the Secretary of Health and Human Services, has determined to permit the word "organic" to appear on the ingredient listing panel to describe those ingredients that are organically produced in accordance with this title.

(d) SMALL FARMER EXEMPTION.-Subsection (a)(1) shall not apply to persons who sell no more than $5,000 annually in value of agricultural products.

Sec. 2107. [7 U.S.C. 6506] General Requirements

(a) IN GENERAL.-A program established under this title shall-

(1) provide that an agricultural product to be sold or labeled as organically produced must-

(A) be produced only on certified organic farms and handled only through certified organic handling operations in accordance with this title; and

(B) be produced and handled in accordance with such program;

(2) require that producers and handlers desiring to participate under such program establish an organic plan under section 2114;

(3) provide for procedures that allow producers and handlers to appeal an adverse administrative determination under this title;

(4) require each certified organic farm or each certified organic handling operation to certify to the Secretary, the governing State official (if applicable), and the certifying agent on an annual basis, that such farm or handler has not produced or handled any agricultural product sold or labeled as organically produced except in accordance with this title;

(5) provide for annual on-site inspection by the certifying agent of each farm and handling operation that has been certified under this title;

(6) require periodic residue testing by certifying agents of agricultural products that have been produced on certified organic farms and handled through certified organic handling operations to determine whether such products contain any pesticide or other nonorganic residue or natural toxicants and to require certifying agents, to the extent that such agents are aware of a violation of applicable laws relating to food safety, to report such violation to the appropriate health agencies;

(7) provide for appropriate and adequate enforcement procedures, as determined by the Secretary to be necessary and consistent with this title;

(8) protect against conflict-of-interest as specified under section 2116(h);

(9) provide for public access to certification documents and laboratory analyses that pertain to certification;

(10) provide for the collection of reasonable fees from producers, certifying agents and handlers who participate in such program; and

(11) require such other terms and conditions as may be determined by the Secretary to be necessary.

(b) DISCRETIONARY REQUIREMENTS.-An organic certification pro-gram established under this title may-

(1) provide for the certification of an entire farm or handling operation or specific fields of a farm or parts of a handling operation if-

(A) in the case of a farm or field, the area to be certified has distinct, defined boundaries and buffer zones separating the land being operated through the use of organic methods from land that is not being operated through the use of such methods;

(B) the operators of such farm or handling operation maintain records of all organic operations separate from records relating to other operations and make such records available at all times for inspection by the Secretary, the certifying agent, and the governing State official; and

(C) appropriate physical facilities, machinery, and management practices are established to prevent the possibility of a mixing of organic and nonorganic products or a penetration of prohibited chemicals or other substances on the certified area; and

(2) provide for reasonable exemptions from specific requirements of this title (except the provisions of section 2112) with respect to agricultural products produced on certified organic farms if such farms are subject to a Federal or State emergency pest or disease treatment program.

(c) WILD SEAFOOD.-

(1) IN GENERAL.-Notwithstanding the requirement of section 2107(a)(1)(A) requiring products be produced only on certified organic farms, the Secretary shall allow, through regulations promulgated after

public notice and opportunity for comment, wild seafood to be certified or labeled as organic.

(2) CONSULTATION AND ACCOMMODATION.-In carrying out paragraph (1), the Secretary shall-

(A) consult with-

(i) the Secretary of Commerce;

(ii) the National Organic Standards Board established under section 2119;

(iii) producers, processors, and sellers; and

(iv) other interested members of the public; and

(B) to the maximum extent practicable, accommodate the unique characteristics of the industries in the United States that harvest and process wild seafood.

(d) STATE PROGRAM.-A State organic certification program approved under this title may contain additional guidelines governing the production or handling of products sold or labeled as organically produced in such State as required in section 2108.

(e) AVAILABILITY OF FEES.-

(1) ACCOUNT.-Fees collected under subsection (a)(10) (including late payment penalties and interest earned from in-vestment of the fees) shall be credited to the account that in-curs the cost of the services provided under this title.

(2) USE.-The collected fees shall be available to the Secretary, without further appropriation or fiscal-year limitation, to pay the expenses of the Secretary incurred in providing accreditation services under this title.

Sec. 2108. [7 U.S.C. 6507] State Organic Certification Program

(a) IN GENERAL.-The governing State official may prepare and submit a plan for the establishment of a State organic certification program to the Secretary for approval. A State organic certification program must meet the requirements of this title to be approved by the Secretary.

(b) ADDITIONAL REQUIREMENTS.-

(1) AUTHORITY.-A State organic certification program established under subsection (a) may contain more restrictive requirements governing the organic certification of farms and handling operations and the production and handling of agricultural products that are to be sold or labeled as organically produced under this title than are contained in the program established by the Secretary.

(2) CONTENT.-Any additional requirements established under paragraph (1) shall-

(A) further the purposes of this title;
(B) not be inconsistent with this title;
(C) not be discriminatory towards agricultural commodities organically produced in other States in accordance with this title; and
(D) not become effective until approved by the Secretary.

(c) REVIEW AND OTHER DETERMINATIONS.-

(1) SUBSEQUENT REVIEW.-The Secretary shall review State organic certification programs not less than once during each 5-year period following the date of the approval of such pro-grams.

(2) CHANGES IN PROGRAM.-The governing State official, prior to implementing any substantive change to programs approved under this subsection, shall submit such change to the Secretary for approval.

(3) TIME FOR DETERMINATION.-The Secretary shall make a determination concerning any plan, proposed change to a pro-gram, or a review of a program not later than 6 months after receipt of such plan, such proposed change, or the initiation of such review.

Sec. 2109. [7 U.S.C. 6508] Prohibited Crop production Practices and Materials

(a) SEED, SEEDLINGS AND PLANTING PRACTICES.-For a farm to be certified under this title, producers on such farm shall not apply materials to, or engage in practices on, seeds or seedlings that are contrary to, or inconsistent with, the applicable organic certification program.

(b) SOIL AMENDMENTS.-For a farm to be certified under this title, producers on such farm shall not-

(1) use any fertilizers containing synthetic ingredients or any commercially blended fertilizers containing materials prohibited under this title or under the applicable State organic certification program; or

(2) use as a source of nitrogen: phosphorous, lime, potash, or any materials that are inconsistent with the applicable organic certification program.

(c) CROP MANAGEMENT.-For a farm to be certified under this title, producers on such farm shall not-

(1) use natural poisons such as arsenic or lead salts that have long-term effects and persist in the environment, as determined by the applicable governing State official or the Secretary;

(2) use plastic mulches, unless such mulches are removed at the end of each growing or harvest season; or

(3) use transplants that are treated with any synthetic or prohibited material.

Sec. 2110. [7 U.S.C. 6509] Animal Production Practices and Materials

(a) IN GENERAL.-Any livestock that is to be slaughtered and sold or labeled as organically produced shall be raised in accordance with this title.

(b) BREEDER STOCK.-Breeder stock may be purchased from any source if such stock is not in the last third of gestation.

(c) PRACTICES.-For a farm to be certified under this title as an organic farm with respect to the livestock produced by such farm, producers on such farm-

(1) shall feed such livestock organically produced feed that meets the requirements of this title;

(2) shall not use the following feed-

(A) plastic pellets for roughage;

(B)manure refeeding; or

(C)feed formulas containing urea; and

(3) shall not use growth promoters and hormones on such livestock, whether implanted, ingested, or injected, including antibiotics and synthetic trace elements used to stimulate growth or production of such livestock.

(d) HEALTH CARE.-

(1) PROHIBITED PRACTICES.-For a farm to be certified under this title as an organic farm with respect to the livestock produced by such farm, producers on such farm shall not-

(A) use subtherapeutic doses of antibiotics;

(B) use synthetic internal parasiticides on a routine basis; or

(C) administer medication, other than vaccinations, in the absence of illness.

(2) STANDARDS.-The National Organic Standards Board shall recommend to the Secretary standards in addition to those in paragraph (1) for the care of livestock to ensure that such livestock is organically produced.

(e) ADDITIONAL GUIDELINES.-

(1) POULTRY.-With the exception of day old poultry, all poultry from which meat or eggs will be sold or labeled as organically produced shall be raised and handled in accordance with this title prior to and during the period in which such meat or eggs are sold.

(2) DAIRY LIVESTOCK.-

(A) IN GENERAL.-Except as provided in subparagraph (B), a dairy animal from which milk or milk products will be sold or labeled as organically produced shall be raised and handled in accordance with this title for not less than the 12-month period immediately prior to the sale of such milk and milk products.

(B) TRANSITION GUIDELINE.-Crops and forage from land included in the organic system plan of a dairy farm that is in the third year of organic management may be consumed by the dairy animals of the farm during the 12-month period immediately prior to the sale of organic milk and milk products.

(f) LIVESTOCK IDENTIFICATION.-

(1) IN GENERAL.-For a farm to be certified under this title as an organic farm with respect to the livestock produced by such farm, producers on such farm shall keep adequate records and maintain a detailed, verifiable audit trail so that each animal (or in the case of poultry, each flock) can be traced back to such farm.

(2) RECORDS.-In order to carry out paragraph (1), each producer shall keep accurate records on each animal (or in the case of poultry, each flock) including-

(A) amounts and sources of all medications administered; and

(B) all feeds and feed supplements bought and fed.

(g) NOTICE AND PUBLIC COMMENT.-The Secretary shall hold public hearings and shall develop detailed regulations, with notice and public comment, to guide the implementation of the standards for livestock products provided under this section.

Sec. 2111. [7 U.S.C. 6510] Handling

(a) IN GENERAL.-For a handling operation to be certified under this title, each person on such handling operation shall not, with respect to any agricultural product covered by this title-

(1) add any synthetic ingredient not appearing on the National List during the processing or any postharvest handling of the product;

(2) add any ingredient known to contain levels of nitrates, heavy metals, or toxic residues in excess of those permitted by the applicable organic certification program;

(3) add any sulfites, except in the production of wine, nitrates, or nitrites;

(4) add any ingredients that are not organically produced in accordance with this title and the applicable organic certification program, unless

such ingredients are included on the National List and represent not more than 5 percent of the weight of the total finished product (excluding salt and water);

(5) use any packaging materials, storage containers or bins that contain synthetic fungicides, preservatives, or fumigants;

(6) use any bag or container that had previously been in con-tact with any substance in such a manner as to compromise the organic quality of such product; or

(7) use, in such product water that does not meet all Safe Drinking Water Act requirements.

(b) MEAT.-For a farm or handling operation to be organically certified under this title, producers on such farm or persons on such handling operation shall ensure that organically produced meat does not come in contact with nonorganically produced meat.

Sec. 2112. [7 U.S.C. 6511] Additional Guidelines

(a) IN GENERAL.-The Secretary, the applicable governing State official, and the certifying agent shall utilize a system of residue testing to test products sold or labeled as organically produced under this title to assist in the enforcement of this title.

(b) PREHARVEST TESTING.-The Secretary, the applicable governing State official, or the certifying agent may require preharvest tissue testing of any crop grown on soil suspected of harboring contaminants.

(c) COMPLIANCE REVIEW.-

(1) INSPECTION.-If the Secretary, the applicable governing State official, or the certifying agent determines that an agricultural product sold or labeled as organically produced under this title contains any detectable pesticide or other non-organic residue or prohibited natural substance the Secretary, the applicable governing State official, or the certifying agent shall conduct an investigation to determine if the organic certification program has been violated, and may require the producer or handler of such product to prove that any prohibited substance was not applied to such product.

(2) REMOVAL OF ORGANIC LABEL.-If, as determined by the Secretary, the applicable governing State official, or the certifying agent, the investigation conducted under paragraph (1) indicates that the residue is-

(A) the result of intentional application of a prohibited substance; or

(B) present at levels that are greater than unavoidable residual environmental contamination as prescribed by the Secretary or the applicable governing State official in consultation with the appropriate environmental regulatory agencies; such agricultural product shall not be sold or labeled as organically produced under this title.

(d) RECORDKEEPING REQUIREMENTS.-Producers who operate a certified organic farm or handling operation under this title shall maintain records for 5 years concerning the production or handling of agricultural products sold or labeled as organically produced under this title, including-

(1) a detailed history of substances applied to fields or agricultural products; and

(2) the names and addresses of persons who applied such substances, the dates, the rate, and method of application of such substances.

Sec. 2113. [7 U.S.C. 6512] Other Production and Handling Practices

If a production or handling practice is not prohibited or otherwise restricted under this title, such practice shall be permitted unless it is determined that such practice would be inconsistent with the applicable organic certification program.

Sec. 2114. [7 U.S.C. 6513] Organic Plan

(a) IN G E N E . - A producer or handler seeking certification under this title shall submit an organic plan to the certifying agent and the State organic certification program (if applicable), and such plan shall be reviewed by the certifying agent who shall determine if such plan meets the requirements of the programs.

(b) CROP PRODUCTION FARM PLAN.-

(1) SOIL FERTILITY.-An organic plan shall contain provisions designed to foster soil fertility, primarily through the management of the organic content of the soil through proper tillage, crop rotation, and manuring.

(2) MANURING.-

(A) INCLUSION IN ORGANIC PLAN.-An organic plan shall contain terms and conditions that regulate the application of manure to crops.

(B) APPLICATION OF MANURE.-Such organic plan may provide for the application of raw manure only to-

(i) any green manure crop;

(ii) any perennial crop;

(iii) any crop not for human consumption; and

(iv) any crop for human consumption, if such crop is harvested after a reasonable period of time determined by the certifying agent to ensure the safety of such crop, after the most recent application of raw manure, but in no event shall such period be less than 60 days after such application.

(C) CONTAMINATION BY MANURE.-Such organic plan shall prohibit raw manure from being applied to any crop in a way that significantly contributes to water contamination by nitrates or bacteria.

(c) LIVESTOCK PLAN. A n organic livestock plan shall contain provisions designed to foster the organic production of livestock consistent with the purposes of this title.

(d) MIXED CROP LIVESTOCK PRODUCTION.-An organic plan may encompass both the crop production and livestock production requirements in subsections (b) and (c) if both activities are con-ducted by the same producer.

(e) HANDLING PLAN.-An organic handling plan shall contain pro-visions designed to ensure that agricultural products that are sold or labeled as organically produced are produced and handled in a manner that is consistent with the purposes of this title.

(f) MANAGEMENT OF WILD CROPS.-An organic plan for the harvesting of wild crops shall-

(1) designate the area from which the wild crop will be gathered or harvested;

(2) include a 3 year history of the management of the area showing that no prohibited substances have been applied;

(3) include a plan for the harvesting or gathering of the wild crops assuring that such harvesting or gathering will not be destructive to the environment and will sustain the growth and production of the wild crop; and

(4) include provisions that no prohibited substances will be applied by the producer.

(g) LIMITATION ON CONTENT OF PLAN. A n organic plan shall not include any production or handling practices that are inconsistent with this title.

Sec. 2115. [7 U.S.C. 6514] Accreditation Program

(a) IN GENERAL.-The Secretary shall establish and implement a program to accredit a governing State official, and any private per-son, that meets the requirements of this section as a certifying agent for the purpose of certifying a farm or handling operation as a certified organic farm or handling operation.

(b) REQUIREMENTS.-To be accredited as a certifying agent under this section, a governing State official or private person shall-

(1) prepare and submit, to the Secretary, an application for such accreditation;

(2) have sufficient expertise in organic farming and handling techniques as determined by the Secretary; and

(3) comply with the requirements of this section and section 2116.

(c) DURATION OF DESIGNATION.- n accreditation made under this section shall be for a period of not to exceed 5 years, as deter-mined appropriate by the Secretary, and may be renewed.

Sec. 2116. [7 U.S.C. 6515] Requirements of Certifying Agents

(a) ABILITY To IMPLEMENT REQUIREMENTS.-TO be accredited as a certifying agent under section 2115, a governing State official or a person shall be able to fully implement the applicable organic certification program established under this title.

(b) INSPECTORS.-Any certifying agent shall employ a sufficient number of inspectors to implement the applicable organic certification program established under this title, as determined by the Secretary.

(c) IZECORDKEEPING.-

(1) MAINTENANCE OF RECORDS. A n y certifying agent shall maintain all records concerning its activities under this title for a period of not less than 10 years.

(2) ACCESS FOR SECRETARY.-Any certifying agent shall allow representatives of the Secretary and the governing State official access to any and all records concerning the certifying agent's activities under this title.

(3) TRANSFERENCE OF RECORDS.-If any private person that was certified under this title is dissolved or loses its accreditation, all records or copies of records concerning such person's activities under this title shall be transferred to the Secretary and made available to the applicable governing State official.

(d) AGREEMENT.-Any certifying agent shall enter into an agreement with the Secretary under which such agent shall-

(1) agree to carry out the provisions of this title; and

(2) agree to such other terms and conditions as the Secretary determines appropriate.

(e) PRIVATE CERTIFYING AGENT AGREEMENT.-Any certifying agent that is a private person shall, in addition to the agreement required in subsection (d)-

(1) agree to hold the Secretary harmless for any failure on the part of the certifying agent to carry out the provisions of this title; and

(2) furnish reasonable security, in an amount determined by the Secretary, for the purpose of protecting the rights of participants in the applicable organic certification program established under this title.

(f) COMPLIANCE WITH PROGRAM.-Any certifying agent shall fully comply with the terms and conditions of the applicable organic certification program implemented under this title.

(g) CONFIDENTIALITY.-Except as provided in section 2107(a)(9), any certifying agent shall maintain strict confidentiality with respect to its clients under the applicable organic certification pro-gram and may not disclose to third parties (with the exception of the Secretary or the applicable governing State official) any business related information concerning such client obtained while implementing this title.

(h) CONFLICT OF INTEREST. A n y certifying agent shall not-

(1) carry out any inspections of any operation in which such certifying agent, or employee of such certifying agent has, or has had, a commercial interest, including the provision of consultancy services;

(2) accept payment, gifts, or favors of any kind from the business inspected other than prescribed fees; or

(3) provide advice concerning organic practices or techniques for a fee, other than fees established under such program.

(i) ADMINISTRATOR.A certifying agent that is a private person shall nominate the individual who controls the day-to-day operation of the agent.

(j) LOSS OF ACCREDITATION.-

(1) NONCOMPLIANCE.-If the Secretary or the governing State official (if applicable) determines that a certifying agent is not properly adhering to the provisions of this title, the Secretary or such governing State official may suspend such certifying agent's accreditation.

(2) EFFECT ON CERTIFIED OPERATIONS.-If the accreditation of a certifying agent is suspended under paragraph (1), the Secretary or the governing State official (if applicable) shall promptly determine whether farming or handling operations certified by such certifying agent may retain their organic certification.

Sec. 2117. [7 U.S.C. 6516] Peer Review of Certifying Agents

(a) PEER REVIEW.-In determining whether to approve an application for accreditation submitted under section 2115, the Secretary shall consider a report concerning such applicant that shall be prepared by a peer review panel established under subsection (b).

(b) PEER REVIEW PANEL.-TO assist the Secretary in evaluating applications under section 2115, the Secretary may establish a panel of not less than three persons who have expertise in organic farming and handling methods, to evaluate the State governing official or private person that is seeking accreditation as a certifying agent under such section. Not less than two members of such panel shall be persons who are not employees of the Department of Agriculture or of the applicable State government.

Sec. 2118. [7 U.S.C. 6517] National List

(a) IN GENERAL.-The Secretary shall establish a National List of approved and prohibited substances that shall be included in the standards for organic production and handling established under this title in order for such products to be sold or labeled as organically produced under this title.

(b) CONTENT OF LIST.-The list established under subsection (a) shall contain an itemization, by specific use or application, of each synthetic substance permitted under subsection (c)(1) or each natural substance prohibited under subsection (c)(2).

(c) GUIDELINES FOR PROHIBITIONS OR EXEMPTIONS.-

(1) EXEMPTION FOR PROHIBITED SUBSTANCES IN ORGANIC PRODUCTION AND HANDLING OPERATIONS.-The National List may provide for the use of substances in an organic farming or handling operation that are otherwise prohibited under this title only if-

(A) the Secretary determines, in consultation with the Secretary of Health and Human Services and the Administrator of the Environmental Protection Agency, that the use of such substances-

(i) would not be harmful to human health or the environment;

(ii) is necessary to the production or handling of the agricultural product because of the unavailability of wholly natural substitute products; and

(iii) is consistent with organic farming and handling;

(B) the substance-

(i) is used in production and contains an active synthetic ingredient in the following categories: copper and sulfur compounds; toxins derived from bacteria; pheromones, soaps, horticultural oils, fish emulsions, treated seed, vitamins and minerals; livestock parasiticides and medicines and production aids including netting, tree wraps and seals, insect traps, sticky barriers, row covers, and equipment cleansers; or

(ii) is used in production and contains synthetic inert ingredients that are not classified by the Administrator of the Environmental Protection Agency as inerts of toxicological concern; and

(C) the specific exemption is developed using the procedures described in subsection (d).

(2) PROHIBITION ON THE USE OF SPECIFIC NATURAL SUBSTANCES.-The National List may prohibit the use of specific natural substances in an organic farming or handling operation that are otherwise allowed under this title only if-

(A) the Secretary determines, in consultation with the Secretary of Health and Human Services and the Administrator of the Environmental Protection Agency, that the use of such substances-

(i) would be harmful to human health or the environment; and

(ii) is inconsistent with organic farming or handling, and the purposes of this title; and

(B) the specific prohibition is developed using the procedures specified in subsection (d).

(d) PROCEDURE FOR ESTABLISHING NATIONAL LIST.-

(1) IN GENERAL.-The National List established by the Secretary shall be based upon a proposed national list or proposed amendments to the National List developed by the National Organic Standards Board.

(2) No ADDITIONS.-The Secretary may not include exemptions for the use of specific synthetic substances in the National List other than those

exemptions contained in the Pro-posed National List or Proposed Amendments to the National List.

(3) PROHIBITED SUBSTANCES.-In no instance shall the National List include any substance, the presence of which in food has been prohibited by Federal regulatory action.

(4) NOTICE AND COMMENT.-Before establishing the National List or before making any amendments to the National List, the Secretary shall publish the Proposed National List or any Proposed Amendments to the National List in the Federal Register and seek public comment on such proposals. The Secretary shall include in such Notice any changes to such pro-posed list or amendments recommended by the Secretary.

(5) PUBLICATION OF NATIONAL LIST. After evaluating all comments received concerning the Proposed National List or Proposed Amendments to the National List, the Secretary shall publish the final National List in the Federal Register, along with a discussion of comments received.

(6) EXPEDITED PETITIONS FOR COMMERCIALLY UNAVAILABLE ORGANIC AGRICULTURAL PRODUCTS CONSTITUTING LESS THAN 5 PERCENT OF AN ORGANIC PROCESSED PRODUCT.-The Secretary may develop emergency procedures for designating agricultural products that are commercially unavailable in organic form for placement on the National List for a period of time not to exceed 12 months.

(e) SUNSET PROVISION.-No exemption or prohibition contained in the National List shall be valid unless the National Organic Standards Board has reviewed such exemption or prohibition as provided in this section within 5 years of such exemption or prohibition being adopted or reviewed and the Secretary has renewed such exemption or prohibition.

Sec. 2119. [7 U.S.C. 6518] National Organic Standards Board

(a) IN GENERAL.-The Secretary shall establish a National Organic Standards Board (in accordance with the Federal Advisory Committee Act (5 U.S.C. App. 2 et seq.)) (hereafter referred to in this section as the "Board") to assist in the development of standards for substances to be used in organic production and to advise the Secretary on any other aspects of the implementation of this title.

(b) COMPOSITION OF BOARD.-The Board shall be composed of 15 members, of which-

(1) four shall be individuals who own or operate an organic farming operation;

(2) two shall be individuals who own or operate an organic handling operation;

(3) one shall be an individual who owns or operates a retail establishment with significant trade in organic products;

(4) three shall be individuals with expertise in areas of environmental protection and resource conservation;

(5) three shall be individuals who represent public interest or consumer interest groups;

(6) one shall be an individual with expertise in the fields of toxicology, ecology, or biochemistry; and

(7) one shall be an individual who is a certifying agent as identified under section 2116.

(c) APPOINTMENT.-Not later than 180 days after the date of enactment of this title, the Secretary shall appoint the members of the Board under paragraph (1) through (6) of subsection (b) (and under subsection (b)(7) at an appropriate date after the certification of individuals as certifying agents under section 2116) from nominations received from organic certifying organizations, States, and other interested persons and organizations.

(d) TERM.-A member of the Board shall serve for a term of 5 years, except that the Secretary shall appoint the original members of the Board for staggered terms. A member cannot serve consecutive terms unless such member served an original term that was less than 5 years.

(e) MEETINGS.-The Secretary shall convene a meeting of the Board not later than 60 days after the appointment of its members and shall convene subsequent meetings on a periodic basis.

(f) COMPENSATION AND EXPENSES.A member of the Board shall serve without compensation. While away from their homes or regular places of business on the business of the Board, members of the Board may be allowed travel expenses, including per diem in lieu of subsistence, as is authorized under section 5703 of title 5, United States Code, for persons employed intermittently in the Government service.

(g) CHAIRPERSON.-The Board shall select a Chairperson for the Board.

(h) QUORUM. A majority of the members of the Board shall constitute a quorum for the purpose of conducting business.

(i) DECISIVE VOTES.-Two-thirds of the votes cast at a meeting of the Board at which a quorum is present shall be decisive of any motion.

(j) OTHER TERMS AND CONDITIONS.-The Secretary shall authorize the Board to hire a staff director and shall detail staff of the Department of Agriculture or allow for the hiring of staff and may, subject to necessary appropriations, pay necessary expenses incurred by such Board in carrying out the provisions of this title, as determined appropriate by the Secretary.

(k) RESPONSIBILITIES OF THE BOARD.-

(1) IN GENERAL.-The Board shall provide recommendations to the Secretary regarding the implementation of this title.

(2) NATIONAL LIST.-The Board shall develop the proposed National List or proposed amendments to the National List for submission to the Secretary in accordance with section 2118.

(3) TECHNICAL ADVISORY PANELS.-The Board shall convene technical advisory panels to provide scientific evaluation of the materials considered for inclusion in the National List. Such panels may include experts in agronomy, entomology, health sciences and other relevant disciplines.

(4) SPECIAL REVIEW OF BOTANICAL PESTICIDES.-The Board shall, prior to the establishment of the National List, review all botanical pesticides used in agricultural production and consider whether any such botanical pesticide should be included in the list of prohibited natural substances.

(5) PRODUCT RESIDUE TESTING.-The Board shall advise the Secretary concerning the testing of organically produced agricultural products for residues caused by unavoidable residual environmental contamination.

(6) EMERGENCY SPRAY PROGRAMS.-The Board shall advise the Secretary concerning rules for exemptions from specific requirements of this title (except the provisions of section 2112) with respect to agricultural products produced on certified organic farms if such farms are subject to a Federal or State emergency pest or disease treatment program.

(1) REQUIREMENTS.-In establishing the proposed National List or proposed amendments to the National List, the Board shall-

(1) review available information from the Environmental Protection Agency, the National Institute of Environmental Health Studies, and such other sources as appropriate, concerning the potential for adverse human and environmental effects of substances considered for inclusion in the proposed National List;

(2) work with manufacturers of substances considered for inclusion in the proposed National List to obtain a complete list of ingredients and determine whether such substances contain inert materials that are synthetically produced; and

(3) submit to the Secretary, along with the proposed National List or any proposed amendments to such list, the results of the Board's evaluation and the evaluation of the technical advisory panel of all substances considered for inclusion in the National List.

(m) EVALUATION.-In evaluating substances considered for inclusion in the proposed National List or proposed amendment to the National List, the Board shall consider-

(1) the potential of such substances for detrimental chemical interactions with other materials used in organic farming systems;

(2) the toxicity and mode of action of the substance and of its breakdown products or any contaminants, and their persistence and areas of concentration in the environment;

(3) the probability of environmental contamination during manufacture, use, misuse or disposal of such substance;

(4) the effect of the substance on human health;

(5) the effects of the substance on biological and chemical interactions in the agroecosystem, including the physiological effects of the substance on soil organisms (including the salt index and solubility of the soil), crops and livestock;

(6) the alternatives to using the substance in terms of practices or other available materials; and

(7) its compatibility with a system of sustainable agriculture.

(n) PETITIONS.-The Board shall establish procedures under which persons may petition the Board for the purpose of evaluating substances for inclusion on the National List.

(o) CONFIDENTIALITY.-Any confidential business information obtained by the Board in carrying out this section shall not be re-leased to the public.

Sec. 2120. [7 U.S.C. 6519] Violations of Title

(a) MISUSE OF LABEL. A n y person who knowingly sells or labels a product as organic, except in accordance with this title, shall be subject to a civil penalty of not more than $10,000.

(b) FALSE STATEMENT. A n y person who makes a false statement under this title to the Secretary, a governing State official, or a certifying agent shall be subject to the provisions of section 1001 of title 18, United States Code.

(c) INELIGIBILITY.-

(1) IN GENERAL.-Except as provided in paragraph (2), any person who-

(A) makes a false statement;

(B) attempts to have a label indicating that an agricultural product is organically produced affixed to such product that such person knows, or should have reason to know, to have been produced or handled in a manner that is not in accordance with this title; or

(C) otherwise violates the purposes of the applicable organic certification program as determined by the Secretary;

after notice and an opportunity to be heard, shall not be eligible, for a period of 5 years from the date of such occurrence, to receive certification under this title with respect to any farm or handling operation in which such person has an interest.

(2) WArvER.-Notwithstanding paragraph (1), the Secretary may reduce or eliminate the period of ineligibility referred to in such paragraph if the Secretary determines that such modification or waiver is in the best interests of the applicable organic certification program established under this title.

(d) REPORTING OF VIOLATIONS.-A certifying agent shall immediately report any violations of this title to the Secretary or the governing State official (if applicable).

(e) VIOLATIONS BY CERTIFYING AGENT.-A certifying agent that is a private person that violates the provisions of this title or that falsely or negligently certifies any farming or handling operation that does not meet the terms and conditions of the applicable organic certification program as an organic operation, as determined by the Secretary or the governing State official (if applicable) shall, after notice and an opportunity to be heard-

(1) lose its accreditation as a certifying agent under this title; and

(2) be ineligible to be accredited as a certifying agent under this title for a period of not less than 3 years subsequent to the date of such determination.

(f) EFFECT OF OTHER LAws.-Nothing in this title shall alter the authority of the Secretary under the Federal Meat Inspection Act (21 U.S.C. 601 et seq.), the Poultry Products Inspection Act (21 U.S.C. 451 et seq.), and the Egg Products Inspection Act (21 U.S.C. 1031 et seq.) concerning meat, poultry and egg products, nor any of the authorities of the Secretary of Health and Human Services under the Federal Food, Drug and Cosmetic Act (21 U.S.C. 301 et seq.), nor the authority of the Administrator of the Environmental Protection Agency under the Federal Insecticide, Fungicide and Rodenticide Act (7 U.S.C. 136 et seq.).

Sec. 2121. [7 U.S.C. 6520] Administrative Appeal

(a) EXPEDITED APPEALS PROCEDURE.-The Secretary shall establish an expedited administrative appeals procedure under which persons may appeal an action of the Secretary, the applicable governing State official, or a certifying agent under this title that-

(1) adversely affects such person; or

(2) is inconsistent with the organic certification program established under this title.

(b) APPEAL OF FINAL DECISION.-A final decision of the Secretary under subsection (a) may be appealed to the United States district court for the district in which such person is located.

Sec. 2122. [7 U.S.C. 6521] Administration

(a) REGULATIONS.-Not later than 540 days after the date of enactment of this title, the Secretary shall issue proposed regulations to carry out this title.

(b) ASSISTANCE TO STATE.-

(1) TECHNICAL AND OTHER ASSISTANCE.-The Secretary shall provide technical, administrative, and Extension Service assistance to assist States in the implementation of an organic certification program under this title.

(2) FINANCIAL ASSISTANCE.-The Secretary may provide financial assistance to any State that implements an organic certification program under this title.

Sec. 2123. [7 U.S.C. 6522] Authorization of Appropriations

There are authorized to be appropriated for each fiscal year such sums as may be necessary to carry out this title.

In: Organic Agriculture in the U.S.
Editor: Alison J. Wellson, pp. 223-247
ISBN: 978-1-60021-305-2

Chapter 7

ORGANIC FARMING AND MARKETING: PUBLICATIONS FROM THE UNITED STATES DEPARTMENT OF AGRICULTURE, 1977-2005*

Mary V. Gold†
Alternative Farming Systems Information Center, National Agricultural Library, Agricultural Research Service, U.S. Department of Agriculture, 10301 Baltimore Avenue, Room 132, Beltsville, MD 20705-2351

This compilation cites only documents published directly by the U.S. Department of Agriculture's primary agencies during the years 1977 through August 2005. The full range of published work by USDA scientists, analysts, advisors and other employees at stations throughout the world appears in many other publishing venues including peer-reviewed journals, research reports, newsletters, conference presentations, Web pages, bulletins and factsheets, databases and more. USDA funded projects and organizations also publish documents that appear in many guises and places.

---

* Note: This bibliography expands and updates the previously published, Organic Farming and Marketing: Publications from USDA, 1993-2003. Citations are arranged by agency name, and then by publication date, earliest to most recent. Over the years, agency names have changed; publications published under the auspices of older agency entities are placed with their current agency counterpart. Many publications are cited with current Internet access information; each URL was verified as correct September 8, 2005. National Agricultural Library call numbers are included when available

† E-mail address: mgold@nal.usda.gov; 301-504-6559;

In terms of subject, only publications pertinent to organic production, food and marketing are represented. There are many others that are relevant to organic systems that are not cited. In addition, this listing does not address the large body of research published before 1940 when reliance on synthetic fertilizers and pesticides in agriculture began to become widely adopted. Much of this historic information, which predates the common use of the term "organic agriculture," is directly applicable to current organic systems. For information on other organic-related resources from USDA and elsewhere, or for information about accessing materials from the National Agricultural Library collection, contact the Alternative Farming Systems Information Center (AFSIC).

Agricultural Marketing Service (AMS)
Agricultural Marketing Service (AMS), National Organic Program (NOP)
Agricultural Research Service (ARS)
Agricultural Research Service (ARS), National Agricultural Library (NAL)
Cooperative State Research, Education, and Extension Service (CSREES)
Cooperative State Research, Education, and Extension Service (CSREES), Sustainable Agriculture Research and Education (SARE)
Economic Research Service (ERS)
Foreign Agricultural Service (FAS)
Natural Resources Conservation Service (NRCS)
Risk Management Agency (RMA)
Rural Development (RD)
World Agricultural Outlook Board (WAOB)
Other USDA Publications

Agricultural Marketing Service (AMS)

Organic Food and Fiber: An Analysis of 1994 Certified Production in the United States, by Julie Anton Dunn. USDA, Agricultural Marketing Service, Transportation and Marketing Division, 1995. 5p. [NAL Call Number: aS441.D86 1995]

Agricultural Marketing Service (AMS), National Organic Program (NOP)

Labeling Alcoholic Beverage Containers. USDA, Agricultural Marketing Service, National Organic Program, 2002.
Full text online: http://www.ams.usda.gov/nop/ProdHandlers/AlcoholLabel.htm

Organic Food Standards and Labels: The Facts. USDA, Agricultural Marketing Service, National Organic Program, 2002.
Full text online: http://www.ams.usda.gov/nop/Consumers/brochure.html

How Retail Food Establishments Can Comply with National Organic Program Regulations. USDA, Agricultural Marketing Service, National Organic Program, 2003.
Full text online:
http://www.ams.usda.gov/nop/ProdHandlers/RetailFoodEstablishments.html

Labeling Packaged Products. USDA, Agricultural Marketing Service, National Organic Program, 2003.
Full text online: http://www.ams.usda.gov/nop/ProdHandlers/LabelTable.htm

Organic Feed for Poultry and Livestock: Availability and Prices. USDA, Agricultural Marketing Service, National Organic Program, 2003.
Full text online:
http://www.ams.usda.gov/nop/ProdHandlers/FeedStudyJune2003.pdf

Value Through Verification: USDA National Organic Program, by Barbara Robinson. USDA, Agricultural Marketing Service, National Organic Program, 2004. Note: Available in both PowerPoint and HTML formats.
Full text online: http://www.ams.usda.gov/nop/NOPPresentation/home.html

Accredited Certifying Agents. USDA, Agricultural Marketing Service, National Organic Program, 2005.
Full text online: http://www.ams.usda.gov/nop/CertifyingAgents/Accredited.html

Questions and Answers. USDA, Agricultural Marketing Service, National Organic Program, updated monthly.
Full text online: http://www.ams.usda.gov/nop/Q&A.html

AGRICULTURAL RESEARCH SERVICE (ARS)

A Bibliography for Small and Organic Farmers, 1920 to 1978, by James W. Schwartz. USDA, Science and Education Administration, Agricultural Research, 1978. 157p. [NAL Call Number: aZ5074.O7S3]

Description: "The bibliography presented here consists of 1,176 publications of long-term research by scientists of the Science and Education Administration - Agricultural Research's Soil, Water, and Air Sciences staff (SWAS) from 1920-1978 that relate to the needs of both the small and organic farmers. Only those publications that were considered most useful to these groups of farmers were selected. They are listed under 19 subject areas according to the year published. A brief summary of the subject area appears at the beginning of each section." [From the Introduction]

"Postharvest Shelf Life of Organically Grown Produce: Current Perceptions and Need for Further Study in the USA," by T. Schettini. USDA, Agricultural Research Service. In Workshop on Biological Control of Postharvest Diseases of Fruits and Vegetables. (June 1991): 218-225. (ARS, 92) Note: Paper presented at meeting held September 12-14, 1990, Sheperdstown, WV. [NAL Call Number: aS21.R44A7]

Sustainable Production of Fresh-Market Tomatoes and Other Summer Vegetables with Organic Mulches, by Aref A. Abdul-Baki and John R. Teasdale. Rev. edition. (Farmers' Bulletin, 2279) USDA, Agricultural Research Service, Sustainable Agricultural Systems Laboratory (Beltsville MD), August 1997. [NAL Call Number: 1 Ag84F no. 2279]
Full text online: http://www.ars.usda.gov/is/np/tomatoes.html

"Research for MeBr Alternatives May Benefit Organic Growers." USDA, Agricultural Research Service. Methyl Bromide Alternatives 5, no. 4 (October 1999). [NAL Call Number: aSB952.B75M47]
Full text online: http://www.ars.usda.gov/is/np/mba/oct99/organic.htm

"Turning to the Sun—Instead of Methyl Bromide," by Judy McBride. USDA, Agricultural Research Service. Agricultural Research 47, no. 10 (October 1999): 18-19. [NAL Call Number: 1.98 Ag84]
Full text online: http://www.ars.usda.gov/is/AR/archive/oct99/sun1099.htm

Organic Growers Join with ARS to Retain Crop Diversity, by Kathryn Barry Stelljes. (ARS Press Release) USDA, Agricultural Research Service, October 20, 1999.
Full text online: http://www.ars.usda.gov/is/pr/1999/991020.htm

"Fungal Protein Slows Broadleaf Weeds," by Jan Suszkiw. USDA, Agricultural Research Service. Agricultural Research 48, no. 2 (February 2000): 10-11. [NAL Call Number: 1.98 Ag84]
Full text online: http://www.ars.usda.gov/is/AR/archive/feb00/weeds0200.htm

"Soil Fungi Critical to Organic Success," by Don Comis. USDA, Agricultural Research Service. Agricultural Research 49, no. 5 (May 2001): 20-21. [NAL Call Number: 1.98 Ag84]
Full text online: http://www.ars.usda.gov/is/AR/archive/may01/fungi0501.htm

"An Organic Cure for a Turfgrass Disease?" by Ben Hardin. USDA, Agricultural Research Service. Agricultural Research 49, no. 8 (August 2001): 12-13. [NAL Call Number: 1.98 Ag84]
Full text online: http://www.ars.usda.gov/is/AR/archive/aug01/turf0801.htm

The Rodale Institute: USDA ARS - 044 Final Report: August 1997-December 2000. USDA, Agricultural Research Service, 2002? 24p. [NAL Call Number: aS605.5.R63 2002]

"Forum - Adding OOMPH to Organic Farming," by Michael D. Jawson. USDA, Agricultural Research Service. Agricultural Research 50, no. 2 (February 2002). [NAL Call Number: 1.98 Ag84]
Full text online: http://www.ars.usda.gov/is/AR/archive/feb02/form0202.htm

"Organic Grows on America!" by Marcia Wood et al. USDA, Agricultural Research Service. Agricultural Research 50, no. 2 (February 2002): 4-9. [NAL Call Number: 1.98 Ag84]
Full text online: http://www.ars.usda.gov/is/AR/archive/feb02/amer0202.htm

Better Genes, Not Pills, for Healthier Animals, by Don Comis. (ARS Press Release) USDA, Agricultural Research Service, February 22, 2002.
Full text online: http://www.ars.usda.gov/is/pr/2002/020222.htm

Organic Research Matters [periodic newsletter]. USDA, Agricultural Research Service, Organic Research Program (Salinas, CA), 2004 - .
Full text online: http://www.ars.usda.gov/Research/docs.htm?docid=6607

"Soil Fungi Plants' Natural Friend," by David Elstein. USDA, Agricultural Research Service. Agricultural Research 52, no. 5 (May 2004): 7. [NAL Call Number: 1.98 Ag84]
Full text online: http://www.ars.usda.gov/is/AR/archive/may04/fungi0504.htm

Organic Farming is a Winner for Sustainability, by Don Comis. (ARS Press Release) USDA, Agricultural Research Service, November 26, 2004.
Full text online: http://www.ars.usda.gov/is/pr/2004/041126.htm

Agricultural Research Service (ARS), National Agricultural Library (NAL)

Organic Farming and Gardening: 1969-1976 and 1969-1978, by Charles N. Bebee. (Quick Bibliography Series, 77-17; 79-28) USDA, National Agricultural Library, 1977; 1979. [NAL Call Number: aZ5071.N3]

Organic Farming and Gardening: 1970-1988, by Jayne T. MacLean. (Quick Bibliography Series, 80-27; 83-40; 84-42; 85-25; 86-57; 87-10; 88-38) USDA, National Agricultural Library, 1980-1988 . Note: Seven bibliographies in this series, covering 1970-1986. [NAL Call Number: aZ5071.N3]

Educational and Training Opportunities in Organic, Low Input or Sustainable Agriculture. USDA, National Agricultural Library, Alternative Farming Systems Information Center, 1988. 10p. [NAL Call Number: aS605.5.E38]

Tracing the Evolution of Organic/Sustainable Agriculture: A Selected and Annotated Bibliography, by Jane Potter Gates. USDA, National Agricultural Library, Alternative Farming Systems Information Center, 1988. 20p. [NAL Call Number: aZ5076.A1U54 no.72]
Full text online: http://www.nal.usda.gov/afsic/AFSIC_pubs/tracing.htm

Organic, Low-input or Sustainable Agriculture, 1984 -1989, by Jayne T. MacLean. (Quick Bibliography Series, 88-38; 89-15; 89-54) USDA, National Agricultural Library, Alternative Farming Systems Information Center, 1988-

1989. Note: Three bibliographies in this series, covering 1984-March 1989. [NAL Call Number: aZ5071.N3]

Oral History Interview with Mr. Robert Rodale, by Jane Potter Gates and Robert Rodale. USDA, National Agricultural Library, Alternative Farming Systems Information Center, 1989. Note: 1 videocassette (69 min., 36 sec.): sd., col. (VHS). [NAL Call Number: Videocassette no. 670]

Educational and Training Opportunities in Sustainable Agriculture [annual directory]. USDA, National Agricultural Library, Alternative Farming Systems Information Center, 1990- . [NAL Call Number: aS605.5.E38]
Full text online: http://www.nal.usda.gov/afsic/AFSIC_pubs/edtr.htm

Oral History Interview with Fred Kirschenmann, by Jane Potter Gates and Fred Kirschenmann. USDA, National Agricultural Library, Alternative Farming Systems Information Center, 1990. Note: 1 videocassette (53 min.): sd., col. (VHS). [NAL Call Number: Videocassette no. 877]

Oral History Interview with James Patrick Madden, by Jane Potter Gates and J. Patrick Madden. USDA, National Agricultural Library, Alternative Farming Systems Information Center, 1990. Note: 1 videocassette (55 min.): sd., col. (VHS). [NAL Call Number: Videocassette no. 1009]

Organic Certification, by Jane Potter Gates. (Special Reference Briefs, 90-04) USDA, National Agricultural Library, Alternative Farming Systems Information Center, January 1990. 9p. Note: Bibliography [NAL Call Number: aS21.D27S64]

Societal Impacts of Adoption of Alternative Agricultural Practices, January 1979-April 1990, by Jayne T. MacLean. (Quick Bibliography Series, 90-57) USDA, National Agricultural Library, Alternative Farming Systems Information Center, June 1990. 16p. Note: Bibliography. [NAL Call Number: aZ5071.N3]

Oral History Interview with Dick Thompson, by Jane Potter Gates and Dick Thompson. USDA, National Agricultural Library, Alternative Farming Systems Information Center, 1991. Note: 1 videocassette (60 min.): sd., col. (VHS). [NAL Call Number: Videocassette no. 1008]

Oral History Interview with I. Garth Youngberg, by Jane Potter Gates and Garth Youngberg. USDA, National Agricultural Library, Alternative Farming Systems

Information Center, 1991. Note: 1 videocassette (60 min.): sd., col. (VHS). [NAL Call Number: Videocassette no. 1128]

Organic Certification, by Jane Potter Gates. (Special Reference Briefs, 91-10) USDA, National Agricultural Library, Alternative Farming Systems Information Center, August 1991. 18p. [NAL Call Number: aS21.D27S64]

Societal Impacts of Adoption of Alternative Agricultural Practices, January 1985-August 1992, by Jayne T. MacLean. (Quick Bibliography Series, 93-01) USDA, National Agricultural Library, Alternative Farming Systems Information Center, October 1992. [NAL Call Number: aZ5071.N3 no.93-01]
Full text online: http://www.nal.usda.gov/afsic/AFSIC_pubs/qb93-01.htm

Organic Production: Recent Publications and Current Information Sources, by Mary V. Gold. (Special Reference Briefs, 96-07) USDA, National Agricultural Library, Alternative Farming Systems Information Center, 1996. [NAL Call Number: aS21.D27 S64 no. 96-07]
Full text online: http://www.nal.usda.gov/afsic/AFSIC_pubs/srb96-07.htm

Organic Production: Economic Aspects. Books, Articles, and Videocassettes, 1991-March 1997, by Mary V. Gold. USDA, National Agricultural Library, Alternative Farming Systems Information Center, 1997.
Full text online: http://www.nal.usda.gov/afsic/AFSIC_pubs/organic1.htm

Organically Produced Foods: Nutritive Content, by Mary V. Gold. (Special Reference Briefs, 2000-03) USDA, National Agricultural Library, Alternative Farming Systems Information Center, 2000. 21p. [NAL Call Number: aS21.D27 S64 no. 2000-03]
Full text online: http://www.nal.usda.gov/afsic/AFSIC_pubs/srb0003.htm

Organic Foods: Markets and Marketing - Recent Books, Articles, News Items, Reports, and Videos, by Mary V. Gold. USDA, National Agricultural Library, Alternative Farming Systems Information Center, 2001.
Full text online: http://www.nal.usda.gov/afsic/AFSIC_pubs/orgmrkt.htm

Organic Information Resources: What Are They? Where Are They? How Can I Find Them? by Mary V. Gold. USDA, National Agricultural Library, Alternative Farming Systems Information Center, 2001.
Full text online: http://www.nal.usda.gov/afsic/ofp/orgfind.htm

Organic Gardening: A Guide to Resources 1989-September 2003, by Stephanie M. Boehmer. USDA, National Agricultural Library, Alternative Farming Systems Information Center, September 2003.
Full text online: http://www.nal.usda.gov/afsic/AFSIC_pubs/org_gar.htm

Organic Livestock Production: A Bibliography: Selected Citations from 1974-September 2004, by Mary V. Gold. (Special Reference Briefs, 2004-05) USDA, National Agricultural Library, Alternative Farming Systems Information Center, September 2004.
Full text online: http://www.nal.usda.gov/afsic/AFSIC_pubs/srb0405.htm

Organic Aquaculture, by Stephanie Boehmer et al. (AFSIC Notes, 5) USDA, National Agricultural Library, Alternative Farming Systems Information Center, January 2005.
Full text online: http://www.nal.usda.gov/afsic/AFSIC_pubs/afnotes5.htm

Organic Agricultural Products: Marketing and Trade Resources, by Mary V. Gold. Updated edition. (Special Reference Briefs, 2003-01) USDA, National Agricultural Library, Alternative Farming Systems Information Center, September 2005. 142p.
Full text online: http://www.nal.usda.gov/afsic/AFSIC_pubs/OAP/srb0301.htm

Cooperative State Research, Education, and Extension Service (CSREES)

Organic-related Farming Systems Research FY 82: A Directory of USDA and State Projects in CRIS. USDA, Current Research Information System (CRIS), 1985. 459p. [NAL Call Number: aS605.5.O7]

Growth in Organic Markets Helps Launch Florida Research Center. (Selected Results and Impacts) USDA, Cooperative State Research, Education, and Extension Service, July 2004.
Full text online:
http://www.csrees.usda.gov/nea/ag_systems/sri/sustain_ag_sri_sare_growth.html

Second Year for Integrated Organic Program. (In Focus) USDA, Cooperative State Research, Education, and Extension Service, April 2005.

Full text online:
http://www.csrees.usda.gov/nea/ag_systems/in_focus/organic_if_iop.html

COOPERATIVE STATE RESEARCH, EDUCATION, AND EXTENSION SERVICE (CSREES), SUSTAINABLE AGRICULTURE RESEARCH AND EDUCATION (SARE)

The Real Dirt: Farmers Tell about Organic and Low-Input Practices in the Northeast, by Miranda Smith, Members of the Northeast Organic Farming Association, and Cooperative Extension. USDA, Cooperative State Research, Education, and Extension Service, Sustainable Agriculture Research and Education, Northeast Region, 1994. 264p. [NAL Call Number: S605.5.R43 1994]

CSA Farms in the United States 1999-2000: Community Supported Agriculture (CSA): A New Partnership Between Farmers and Consumers. USDA, Cooperative State Research, Education, and Extension Service, Sustainable Agriculture Research and Education/Sustainable Agriculture Network, 1999. 51p. [NAL Call Number: aHD1484 .C73 1999]

Explore Organic Production. (Farming for Profit, Stewardship & Community Tip Sheet, 4) USDA, Cooperative State Research, Education, and Extension Service, Sustainable Agriculture Research and Education/Sustainable Agriculture Network, 1999.

Naturalize Your Farming System: A Whole-Farm Approach to Managing Pests. (Information Bulletin) USDA, Cooperative State Research, Education, and Extension Service, Sustainable Agriculture Research and Education/Sustainable Agriculture Network, 2000. 20p. [NAL Call Number: SB950.W463-2000]
Full text online: http://www.sare.org/publications/farmpest.htm

Profitable Poultry: Raising Birds on Pasture. (Information Bulletin) USDA, Cooperative State Research, Education, and Extension Service, Sustainable Agriculture Research and Education/Sustainable Agriculture Network, 2002. 15p. [NAL Call Number: SF487.8.A1 P76 2002]
Full text online: http://www.sare.org/bulletin/poultry/

Exploring Sustainability in Agriculture. USDA, Cooperative State Research, Education, and Extension Service, Sustainable Agriculture Research and Education/Sustainable Agriculture Network, 2003. 13p. [NAL Call Number: aS441.E98 2003]
Full text online: http://www.sare.org/bulletin/explore/

Transitioning to Organic Production. (Opportunities in Agriculture) USDA, Cooperative State Research, Education, and Extension Service, Sustainable Agriculture Research and Education/Sustainable Agriculture Network, 2003. 32p. [NAL Call Number: S605.5.O673 2003]
Full text online: http://www.sare.org/publications/organic.htm

ECONOMIC RESEARCH SERVICE (ERS)

"Organic Certification Programs," by Charlene C. Price and Judy Brown. USDA, Economic Research Service. National Food Review, no. 15 (Summer 1981): 31-33. [NAL Call Number: aHD9001.N275]

"Organic Farming Grows in Western Europe," by S. Sposato. USDA, Economic Research Service. National Food Review, no. 24 (Fall 1983): 22-23. [NAL Call Number: aHD9001.N275]

"Organically Speaking, Western Europe is Top Seed," by S. Sposato. USDA, Economic Research Service. Farmline 5, no. 4 (April 1984): 16-17. [NAL Call Number: AGE aHD1401.A2U52]

"Natural Foods: What's in a Name," by Charlene C. Price. USDA, Economic Research Service. Farmline 5, no. 11 (November 1984): 14. [NAL Call Number: AGE aHD1401.A2U52]

"Eating 'Natural' Gains Popularity," by Charlene C. Price. USDA, Economic Research Service. National Food Review, no. 28 (1985): 14-18. [NAL Call Number: aHD9001.N275]

Alternative Agriculture: Growing Concerns. USDA, Economic Research Service, 1989. Note: 1 videocassette (19 min.): sd., col. (VHS). [NAL Call Number: Videocassette no. 479]

"The Food Industry Responds to Consumers' Pesticide Fears," by E. Van Ravenswaay. USDA, Economic Research Service, Commodity Economics Division. FoodReview 12, no. 3 (July/Sept 1989): 17-20. [NAL Call Number: aHD9001.N275]

"Success Steady in Organic Produce," by Catherine Greene. USDA, Economic Research Service. Agricultural Outlook, no. 185 (May 1992): 15-17. [NAL Call Number: aHD1751.A42]

"Organic Certification: Standards in the Works," by Julie Anton and Betsy Frazão. USDA, Economic Research Service. Agricultural Outlook, no. 199 (1993): 26-31. [NAL Call Number: aHD1751.A42]

"Some Barriers to Organic Produce at the Wholesale Level," by S. Payson, B. Lin, and J. Wertz. USDA, Economic Research Service, Resource and Technology Division. FoodReview 17, no. 2 (May/Aug 1994): 23-26. [NAL Call Number: aHD9001.N275]

"Organic Foods Find Opportunity in the Natural Food Industry," by Julie Anton Dunn. USDA, Economic Research Service. FoodReview 18, no. 3 (September/December 1995): 7-18. [NAL Call Number: aHD9001.N275]

Organic Vegetable Growers Surveyed in 1994: Updates on Agricultural Resources and Environmental Indicators, by Jorge Fernandez-Cornejo, Renata Penn, and Doris Newton. USDA, Economic Research Service, May 1996. [NAL Call Number: aHD1751.R732]
Full text online: http://www.ers.usda.gov/publications/arei/96upd/upd96-4.pdf

"'Organically Grown' Vegetables: U.S. Acreage and Markets Expand During the 1990's," by Linda Calvin and Catherine Greene. USDA, Economic Research Service. Vegetables and Specialties Situation and Outlook (May 1997). [NAL Call Number: aHD9220.U5S47]
Full text online: http://usda.mannlib.cornell.edu/reports/erssor/specialty/vgs-bb/1997/vgs271f.asc

"Consumers Pay a Premium for Organic Baby Foods," by J. Michael Harris. USDA, Economic Research Service. FoodReview 20, no. 2 (May/Aug 1997): 13-16. [NAL Call Number: aHD9001.N275]

Full text online:
http://www.ers.usda.gov/publications/foodreview/aug1997/may97d.pdf

Organic Fruit Growers Survey: Updates on Agricultural Resources and Environmental Indicators, by Jorge Fernandez-Cornejo, Renata Penn, and Doris Newton. USDA, Economic Research Service, June 1997. [NAL Call Number: aHD1751.R732]
Full text online: http://www.ers.usda.gov/publications/arei/97upd/upd97-4.pdf

"Demand for Frozen Vegetables: A Comparison of Organic and Conventional Products," by Lewrene Glaser et al. USDA, Economic Research Service. Vegetables and Specialities (November 1998). [NAL Call Number: aHD9220.U5S47]
Full text online: http://usda.mannlib.cornell.edu/reports/erssor/specialty/vgs-bb/1998/vgs276f.asc

"U.S. Organic Agriculture Gaining Ground," by Catherine Greene. USDA, Economic Research Service. Agricultural Outlook, no. 270 (April 2000): 9-14. [NAL Call Number: aHD1751.A42]
Full text online: http://www.ers.usda.gov/publications/agoutlook/apr2000/ao270d.pdf

"Organic Foods: Niche Marketers Venture into the Mainstream," by Carolyn Dimitri and Nessa Richman. USDA, Economic Research Service. Agricultural Outlook, no. 272 (June/July 2000): 11-14. [NAL Call Number: aHD1751.A42]
Full text online: http://www.ers.usda.gov/publications/AgOutlook/Jun2000/ao272f.pdf

Economics of Food Labeling, by Elise Golan et al. (Agricultural Economic Report, 793) USDA, Economic Research Service, 2001. vi, 41p. [NAL Call Number: A281.9 Ag8A no. 793]
Full text online: http://www.ers.usda.gov/publications/aer793/

U.S. Organic Farming Emerges in the 1990s: Adoption of Certified Systems, by Catherine Greene. (Agriculture Information Bulletin, 770) USDA, Economic Research Service, 2001. iv, 24p. [NAL Call Number: 1 Ag84Ab no. 770]
Full text online: http://www.ers.usda.gov/publications/aib770/

"Organic Marketing Features Fresh Food and Direct Exchange," by Catherine Greene, Carolyn Dimitri, and Nessa Richman. USDA, Economic Research Service. FoodReview 24, no. 1 (January/April 2001): 31-37. [NAL Call Number: aHD9001.N275]
Full text online: http://www.ers.usda.gov/publications/FoodReview/jan2001/frv24I1f.pdf

"Organic Wheat Production in the United States: Expanding Markets and Supplies," by Catherine Greene and Thomas Dobbs. USDA, Economic Research Service. Wheat Situation and Outlook Yearbook (March 2001).
Full text online: http://www.ers.usda.gov/publications/so/view.asp?f=field/whs-bby/whs2001.pdf

"Tracking Wholesale Prices for Organic Produce," by Emy Sok and Lewrene Glaser. USDA, Economic Research Service. Agricultural Outlook, no. 285 (October 2001): 7-8. [NAL Call Number: aHD1751.A42]
Full text online:
http://www.ers.usda.gov/publications/agoutlook/oct2001/ao285d.pdf

Organic Agriculture Provisions [2002 Farm Bill]. (ERS Analysis) USDA, Economic Research Service, 2002.
Full text online:
http://www.ers.usda.gov/Features/farmbill/analysis/organicagriculture.htm

Recent Growth Patterns in the U.S. Organic Foods Market, by Carolyn Dimitri and Catherine Greene. (Agriculture Information Bulletin, 777) USDA, Economic Research Service, September 2002. iii, 39p. [NAL Call Number: 1 Ag84Ab no. 777]
Full text online: http://www.ers.usda.gov/publications/aib777/

"Organic Food Industry Taps Growing American Market," by Carolyn and Catherine Greene Dimitri. USDA, Economic Research Service. Agricultural Outlook, no. 295 (October 2002): 4-7. [NAL Call Number: aHD1751.A42]
Full text online:
http://www.ers.usda.gov/publications/agoutlook/oct2002/ao295b.pdf

Organic Production [data set]. USDA, Economic Research Service, October 2002.
Full text online: http://www.ers.usda.gov/Data/Organic/

"U.S. Organic Farming: A Decade of Expansion," by Catherine Greene and Amy Kremen. USDA, Economic Research Service. Agricultural Outlook, no. 296 (November 2002): 31-34. [NAL Call Number: aHD1751.A42]
Full text online:
http://www.ers.usda.gov/publications/agoutlook/Nov2002/ao296i.pdf

U.S. Organic Farming in 2000-2001: Adoption of Certified Systems, by Catherine Greene and Amy Kremen. (Agriculture Information Bulletin, 780) USDA, Economic Research Service, 2003. iv, 51p. [NAL Call Number: 1 Ag84Ab no. 780]
Full text online: http://www.ers.usda.gov/publications/aib780/

"Organic Agriculture: Gaining Ground," by Catherine Greene and Carolyn Dimitri. USDA, Economic Research Service. Amber Waves (February 2003). [NAL Call Number: aHD1751.A52]
Full text online: http://www.ers.usda.gov/AmberWaves/Feb03/Findings/Organic Agriculture.htm

Organic Farmgate and Wholesale Prices [data set]. USDA, Economic Research Service, May 2005.
Full text online: http://www.ers.usda.gov/Data/OrganicPrices/

Price Premiums Hold on as U.S. Organic Produce Market Expands, by Lydia Oberholtzer, Carolyn Dimitri, and Catherine Greene. (Outlook Report, VGS30801) USDA, Economic Research Service, May 2005. 22p.
Full text online: http://www.ers.usda.gov/Publications/vgs/may05/VGS30801/

Market-Led Growth vs. Government-Facilitated Growth: Development of the U.S. and EU Organic Agricultural Sectors, by Carolyn Dimitri and Lydia Oberholtzer. (Outlook Report, WRS0505) USDA, Economic Research Service, August 2005. 26p.
Full text online: http://www.ers.usda.gov/Publications/wrs0505/

"Organic Price Premiums Remain High," by Carolyn Dimitri and Lydia Oberholtzer. USDA, Economic Research Service. Amber Waves (September 2005). [NAL Call Number: aHD1751.A52]
Full text online:
http://www.ers.usda.gov/AmberWaves/September05/Findings/OrganicPrice.htm

FOREIGN AGRICULTURAL SERVICE (FAS)

"European Yuppies - The New Niche Market for U.S. Food Exporters." USDA, Foreign Agricultural Service. AgExporter 1, no. 12 (December 1989): 4-8. [NAL Call Number: aHD1401.A7]

"French Organics Market Awaits Development," by R. Gauthier. USDA, Foreign Agricultural Service. AgExporter 8, no. 6 (August/September 1996): 17-19. [NAL Call Number: aHD1401.A7]

"Japan Buys into Organics," by J. K. Jenni. USDA, Foreign Agricultural Service. AgExporter 8, no. 7 (October 1996): 20-22. [NAL Call Number: aHD1401.A7]

"Dutch Organic Food Market Offers All-Natural Potential for U.S. Firms," by Tatiana van der Harst-Collaris and Laura Gabel Scandurra. USDA, Foreign Agricultural Service. AgExporter 9, no. 8 (August 1997): 10-15. [NAL Call Number: aHD1401.A7]
Full text online: http://www.fas.usda.gov/info/agexporter/1997/dutch.html

"Defining Moment: Canadian Organic Regs Are Coming," by George C. Myles. USDA, Foreign Agricultural Service. AgExporter 9, no. 11 (November 1997): 24-26. [NAL Call Number: aHD1401.A7]
Full text online: http://www.fas.usda.gov/info/agexporter/1997/defining.html

Organic Perspectives [quarterly newsletter]. USDA, Foreign Agricultural Service. 1998-.
Full text online: http://www.fas.usda.gov/agx/organics/organics.html

"France's Baby Food Boom," by Mattie R. Sharpless and Roselyne Gauthier. USDA, Foreign Agricultural Service. AgExporter 10, no. 2 (February 1998): 18-19. [NAL Call Number: aHD1401.A7]
Full text online:
http://www.fas.usda.gov/info/agexporter/1998/February%201998/frances.html

"Organic Food and Vegetable Products, a Sino- the Times," by Henry Lin. USDA, Foreign Agricultural Service. AgExporter 10, no. 3 (March 1998): 8-9. [NAL Call Number: aHD1401.A7]
Full text online:
http://www.fas.usda.gov/info/agexporter/1998/March%201998/organic.html

"Organic Food Trend Sprouts in Hong Kong," by W. Garth Thorburn II. USDA, Foreign Agricultural Service. AgExporter 10, no. 9 (September 1998): 15. [NAL Call Number: aHD1401.A7]
Full text online:
http://www.fas.usda.gov/info/agexporter/1998/Sept%201998/orgHoKoc.html

"Organic Food Trend Wins in Taiwan," by Chiou Mey Perng. USDA, Foreign Agricultural Service. AgExporter 10, no. 9 (September 1998): 13-14. [NAL Call Number: aHD1401.A7]
Full text online:
http://www.fas.usda.gov/info/agexporter/1998/Sept%201998/orgtaiw.html

"Singapore: A Growing Market for U.S. Health Foods," by Dale L. Good. USDA, Foreign Agricultural Service. AgExporter 10, no. 9 (September 1998): 10-12. [NAL Call Number: aHD1401.A7]
Full text online:
http://www.fas.usda.gov/info/agexporter/1998/Sept%201998/singapor.html

"Organic and Hispanic, These West Coast Chipsters Are Bagging Exports," by Jill Lee. USDA, Foreign Agricultural Service. AgExporter 11, no. 12 (December 1999): 21. [NAL Call Number: aHD1401.A7]
Full text online: http://www.fas.usda.gov/info/agexporter/1999/organic.html

"Five Tips for Organic Trade Success," by Jill Lee. USDA, Foreign Agricultural Service. AgExporter 12, no. 6 (June 2000): 11-12. [NAL Call Number: aHD1401.A7]
Full text online: http://www.fas.usda.gov/info/agexporter/2000/june/fivetips.htm

"France: More Than a Nice Niche?" by Frank J. Piason. USDA, Foreign Agricultural Service. AgExporter 12, no. 6 (June 2000). [NAL Call Number: aHD1401.A7]
Full text online: http://www.fas.usda.gov/info/agexporter/2000/june/france.htm

"Germany: The Customer-Competitor Duality," by Dietmar Achilles. USDA, Foreign Agricultural Service. AgExporter 12, no. 6 (June 2000): 20. [NAL Call Number: aHD1401.A7]
Full text online: http://www.fas.usda.gov/info/agexporter/2000/june/germany.htm

"Lacking a Domestic Source, Hong Kong Must Import Organics," by Carolyn Yuen. USDA, Foreign Agricultural Service. AgExporter 12, no. 6 (June 2000): 22. [NAL Call Number: aHD1401.A7]
Full text online: http://www.fas.usda.gov/info/agexporter/2000/june/lackinga.htm

"Organic Markets Offer U.S. Agriculture Current and Future Sales Opportunities," by Janise Zygmont. USDA, Foreign Agricultural Service. AgExporter 12, no. 6 (June 2000): 4-7. [NAL Call Number: aHD1401.A7]
Full text online: http://www.fas.usda.gov/info/agexporter/2000/june/organic.htm

"Organic Products Finding a Ready Market on Taiwan," by Chiou Mey Perng. USDA, Foreign Agricultural Service. AgExporter 12, no. 6 (June 2000): 17-19. [NAL Call Number: aHD1401.A7]
Full text online: http://www.fas.usda.gov/info/agexporter/2000/june/organic3.htm

"Organic Trade Association Sees Opportunities in Exports," by Janise Zygmont. USDA, Foreign Agricultural Service. AgExporter 12, no. 6 (June 2000): 8-10. [NAL Call Number: aHD1401.A7]
Full text online: http://www.fas.usda.gov/info/agexporter/2000/june/organic2.htm

"Practical Contacts for Organic Producers and Exporters" USDA, Foreign Agricultural Service. AgExporter 12, no. 6 (June 2000). [NAL Call Number: aHD1401.A7]
Full text online: http://www.fas.usda.gov/info/agexporter/2000/june/practica.htm

"Timeline of Organic Milestones," USDA, Foreign Agricultural Service. AgExporter 12, no. 6 (June 2000). [NAL Call Number: aHD1401.A7]
Full text online: http://www.fas.usda.gov/info/agexporter/2000/june/timeline.htm

"The U.K. Organic Sector: Go for It!" by Jennifer Jones. USDA, Foreign Agricultural Service. AgExporter 12, no. 6 (June 2000): 15-16. [NAL Call Number: aHD1401.A7]
Full text online: http://www.fas.usda.gov/info/agexporter/2000/june/uk.htm

"Wizard's Cauldron Conjures International Markets," by Jill Lee. USDA, Foreign Agricultural Service. AgExporter 12, no. 6 (June 2000): 13-14. [NAL Call Number: aHD1401.A7]
Full text online: http://www.fas.usda.gov/info/agexporter/2000/june/wizards.htm

Danish Organic Dairy Production, by Melissa Schmaedick. (FAS Division Reports) USDA, Foreign Agricultural Service, July 2000.
Full text online: http://www.fas.usda.gov/dlp2/circular/2000/00-07Dairy/danish.html

U.S. Organic Fruit: Export Opportunities and Competition in the International Market, by Janise Zygmont. (FAS Division Reports) USDA, Foreign Agricultural Service, 2001.
Full text online: http://www.fas.usda.gov/agx/organics/speech.htm

"Organics East: Postcards from the Pacific Market," USDA, Foreign Agricultural Service. AgExporter 13, no. 7 (July 2001): 15-17. [NAL Call Number: aHD1401.A7]
Full text online: http://www.fas.usda.gov/info/agexporter/2001/july/p15-17.pdf

"Big Changes for the Netherlands' Organic Industry," by Wendalin Kolkman. USDA, Foreign Agricultural Service. AgExporter 13, no. 12 (December 2001): 16. [NAL Call Number: aHD1401.A7]
Full text online: http://www.fas.usda.gov/info/agexporter/2001/dec/page16.pdf

"Benelux Opting for Organics," by Marcel Pinckaers. USDA, Foreign Agricultural Service. AgExporter 14, no. 11 (November 2002): 8-9. [NAL Call Number: aHD1401.A7]
Full text online: http://www.fas.usda.gov/info/agexporter/2002/November/pgs8-9.pdf

European Union: Organic Products: European Action Plan for Organic Food and Farming, by Hilde Brans. (GAIN Report, E23020) USDA, Foreign Agricultural Service, Global Agriculture Information Network, 2003.
Full text online: http://www.fas.usda.gov/gainfiles/200302/145884678.pdf

Exporting Organic: Are You Ready for the Challenge? by Kelly Strzelecki. USDA, Foreign Agricultural Service, 2003. Note: PowerPoint presentation.
Full text online: http://www.fas.usda.gov/agx/organics/exporting.htm

Organic, Natural and Wellness Products at Natural Products Expo West Anaheim, CA - March 6-9, 2003. USDA, Foreign Agricultural Service in partnership with Department of Commerce, U.S. Commercial Service, 2003.

Full text online:
http://www.fas.usda.gov/agx/organics/BuyerAlert/BAOrganic.htm

United Kingdom: Organic Products: Market Brief 2003, by Phil Bicknell. (GAIN Report, UK3006) USDA, Foreign Agricultural Service, Global Agriculture Information Network, 2003.
Full text online: http://www.fas.usda.gov/gainfiles/200303/145884996.pdf

"Satisfying the Appetite of the Danes," by Hasse Kristensen. USDA, Foreign Agricultural Service. AgExporter 15, no. 6 (June 2003): 10. [NAL Call Number: aHD1401.A7]
Full text online: http://www.fas.usda.gov/info/agexporter/2003/June/pg%2010.pdf

"Popularity of Organic Foods on the Rise in the United Kingdom," by Philip Bicknell. USDA, Foreign Agricultural Service. AgExporter 15, no. 8 (August 2003): 8-9. [NAL Call Number: aHD1401.A7]
Full text online: http://www.fas.usda.gov/info/agexporter/2003/August/pgs%208-9.pdf

"South Koreans Buying Organic," by Young Sook Oh and Susan B. Phillips. USDA, Foreign Agricultural Service. AgExporter 15, no. 8 (August 2003): 6-7. [NAL Call Number: aHD1401.A7]
Full text online: http://www.fas.usda.gov/info/agexporter/2003/August/pgs%206-7.pdf

"Portuguese Looking to Organics," by Maria do Monte Gomes. USDA, Foreign Agricultural Service. AgExporter 15, no. 10 (October 2003): 14. [NAL Call Number: aHD1401.A7]
Full text online:
http://www.fas.usda.gov/info/agexporter/2003/October/pg%2014.pdf

Attaché Reports: Organics [directory]. USDA, Foreign Agricultural Service, 2004.
Full text online: http://www.fas.usda.gov/agx/organics/attache.htm

"Organics in Austria: From Niche to Mainstream," by Roswitha Krautgartner. USDA, Foreign Agricultural Service. AgExporter 16, no. 2 (February 2004): 6-7. [NAL Call Number: aHD1401.A7]

Full text online:
http://www.fas.usda.gov/info/agexporter/2004/February/pgs%206%20-%207.pdf

"Opportunity Knocks for Organics in Singapore," USDA, Foreign Agricultural Service. AgExporter 16, no. 6 (June 2004): 14-15. [NAL Call Number: aHD1401.A7]
Full text online: http://www.fas.usda.gov/info/agexporter/2004/June/pgs%2014-15.pdf

"Organics Earning Respect on Taiwan," by Welly Liu-Chen and Jeff Miller. USDA, Foreign Agricultural Service. AgExporter 16, no. 6 (June 2004): 13. [NAL Call Number: aHD1401.A7]
Full text online: http://www.fas.usda.gov/info/agexporter/2004/June/pg%2013.pdf

U.S. Market Profile for Organic Food Products, by James M. Tringe. USDA, Foreign Agricultural Service, Commodity and Marketing Programs-Processed Products Division, International Strategic Marketing Group, February 2005. 19p.
Full text online:
http://www.fas.usda.gov/agx/organics/USMarketProfileOrganicFoodFeb2005.pdf

Natural Resources Conservation Service (NRCS)

The Knife Roller (Crimper): An Alternative Kill Method for Cover Crops. (Soil Quality/Agronomy Technical Note, 13) USDA, Natural Resources Conservation Service, Soil Quality Institute, September 2002.
Full text online: http://soils.usda.gov/sqi/files/13.pdf

Risk Management Agency (RMA)

2001 Organic Crop Insurance Underwriting Guide. (FCIC-24140 (03-2001)) USDA, Risk Management Agency, Federal Crop Insurance Corporation, Product Development Division, 2001.
Full text online:
http://www.rma.usda.gov/FTP/Publications/directives/24000/pdf/01_24140.pdf

Rural Development (RD)

The State of U.S. Organic Producer Marketing Cooperatives in 1990, by Alan D. Borst. (Working Paper, 91-W1) USDA, Agricultural Cooperative Service, 1990. v, 18p. [NAL Call Number: ArU HD1491.U5W67 no.91-W1]

The State of U.S. Organic Producer Marketing Cooperatives in 1991, by Alan D. Borst. (Staff Report, 91-S7) USDA, Agricultural Cooperative Service, 1991. v, 23p. [NAL Call Number: aHD9005.B67 1991]

"New Model, Old Ways: Farmers Find New Life Serving Burgeoning Organic Markets," by Pamela J. Karg. USDA, Rural Development, Rural Business-Cooperative Service. Rural Cooperatives 67, no. 1 (January 2000). [NAL Call Number: aHD1491.U6R87] 23-27.
Full text online: http://www.rurdev.usda.gov/rbs/pub/jan00/models.htm

"A Fresh Advantage: Co-op Helps Small Farms Market Produce to High-end Restaurants," by Pamela J. Karg. USDA, Rural Development, Rural Business-Cooperative Service. Rural Cooperatives (May/June 2005): 24-26. [NAL Call Number: aHD1491.U6R87]
Full text online: http://www.rurdev.usda.gov/rbs/pub/may05/may05.pdf

"Going with the Grain: New Mexico Organic Wheat Cooperative Provides Lift to Farmers, Rural Economy," by Catherine Merlo. USDA, Rural Development, Rural Business-Cooperative Service. Rural Cooperatives (May/June 2005): 10-14. [NAL Call Number: aHD1491.U6R87]
Full text online: http://www.rurdev.usda.gov/rbs/pub/may05/may05.pdf

"Slice of the Market: Penns Corner, Tuscarora Co-ops Target Growing Restaurant Trade," by Stephen A. Thompson. USDA, Rural Development, Rural Business-Cooperative Service. Rural Cooperatives (May/June 2005): 27-29. [NAL Call Number: aHD1491.U6R87]
Full text online: http://www.rurdev.usda.gov/rbs/pub/may05/may05.pdf

"USDA B&I Program Supports Growth of Organic Cooperatives," by Peter J. Thomas. USDA, Rural Development, Rural Business-Cooperative Service. Rural Cooperatives (May/June 2005): 43. [NAL Call Number: aHD1491.U6R87]
Full text online: http://www.rurdev.usda.gov/rbs/pub/may05/may05.pdf

"Cream of the CROPP: Demand Outstripping Supply as CROPP Organic Dairy Products Go Prime Time," by Dan Campbell. USDA, Rural Development, Rural Business-Cooperative Service. Rural Cooperatives (May/June 2005): 14-19. [NAL Call Number: aHD1491.U6R87]
Full text online: http://www.rurdev.usda.gov/rbs/pub/may05/may05.pdf

World Agricultural Outlook Board (WAOB)

"Cultivating New Ground: Organic Agriculture Arrives at the Outlook," by Mark Lipson. USDA, World Agricultural Outlook Board. Agricultural Outlook Forum '98 Proceedings (1998): 159-160. Note: Meeting held February 23-24, 1998, in Washington, DC. [NAL Call Number: aHD1755.A376]

"The Evolving Organic Marketplace: Executive Summary," by Harvey Hartman. USDA, World Agricultural Outlook Board. Agricultural Outlook Forum '98 Proceedings (1998): 169-174. Note: Meeting held February 23-24, 1998, in Washington, DC. [NAL Call Number: aHD1755.A376]

"Organic Food Marketing Trends," by Gene Kahn. USDA, World Agricultural Outlook Board. Agricultural Outlook Forum '98 Proceedings (1998). Note: Meeting held February 23-24, 1998, in Washington, DC. PowerPoint file. [NAL Call Number: aHD1755.A376]
Full text online: http://www.usda.gov/oce/waob/outlook98/98speeches.htm

"Direct Marketing Options: Farmers Markets, Restaurants, Community Supported Agriculture, and the Organic Alternative," by Steve Gilman. USDA, World Agricultural Outlook Board. Agricultural Outlook Forum 1999 Proceedings (1999): 118-121. Note: Meeting held February 22-23, 1999, in Arlington, VA. [NAL Call Number: aHD1755.A376]
Full text online:
http://www.usda.gov/agency/oce/waob/outlook99/99speeches.htm

"Organics...Yesterday, Today, and Tomorrow," by Larry Jacobs. USDA, World Agricultural Outlook Board. Agricultural Outlook Forum 2000 (2000). Note: Meeting held February 24-25, 2000, in Arlington, VA.
Full text online:
http://www.usda.gov/oce/waob/Archives/2000/speeches/jacobs.txt

"Venture Capital in Organic Marketing," by Brooks Browne. USDA, World Agricultural Outlook Board. Agricultural Outlook Forum 2000 (2000). Note: Meeting held February 24-25, 2000, in Arlington, VA.
Full text online:
http://www.usda.gov/oce/waob/Archives/2000/speeches/brooksbrowne.txt

"Establishing Protocols for Certified Organic and Natural Beef," by Diana S. Endicott. USDA, World Agricultural Outlook Board. Agricultural Outlook Forum 2003 (2003). Note: Meeting held February 21-22, 2003, in Arlington, VA. Large PowerPoint file.
Full text online:
http://www.usda.gov/oce/forum/Archives/2003/Programs/program.htm

"Current Trends in the Retail Market for Organics: A Retailer's Perspective," by Mary Mulry. USDA, World Agricultural Outlook Board. Agricultural Outlook Forum 2004 (2004). Note: Meeting held February 19-20, 2004, in Arlington, VA.

"The Growth of the Organic Market: Producers' Perspectives," by Kevin Edberg. USDA, World Agricultural Outlook Board. Agricultural Outlook Forum 2004 (2004). Note: Meeting held February 19-20, 2004, in Arlington, VA.
Full text online: http://agecon.lib.umn.edu/cgi-bin/pdf_view.pl?paperid=12736&ftype=.pdf

"Organic Consumer Profile," by Laurie Demeritt. USDA, World Agricultural Outlook Board. Agricultural Outlook Forum 2004 (2004). Note: Meeting held February 19-20, 2004, in Arlington, VA.

"Recent Trends in Organic Production," by Catherine Greene. USDA, World Agricultural Outlook Board. Agricultural Outlook Forum 2004 (2004). Note: Meeting held February 19-20, 2004, in Arlington, VA.
Full text online:
http://www.usda.gov/oce/forum/Archives/2004/speeches/catherinegreene.doc

Other USDA Publications

"How You Can Grow Food Organically," by W. Judkins and F. Smith. In Yearbook of Agriculture. USDA, 1978. 224-229p. [NAL Call Number: 1 AG84Y]

Report and Recommendations on Organic Farming. USDA, Study Team on Organic Farming, 1980. xiv, 94p. [NAL Call Number: aS605.5.U52]

"Cliff and Debbie Van Till: A California Family Finds out What 'Organic' Really Means," by J. Stumbos. In Yearbook of Agriculture, 28-30. (Americans in Agriculture: Portraits of Diversity, edited by D. Smith) USDA, 1990. [NAL Call Number: 1 Ag84y]

First International Conference on Kyusei Nature Farming: Proceedings of the Conference at Khon Kaen University, Khon Kaen, Thailand, October 17-21, 1989, by James F. Parr, S. B. Hornick, and C. E. Whitman. USDA, 1991. ix, 175p. [NAL Call Number: aS605.5.I56-1989]

About the Alternative Farming Systems Information Center at the National Agricultural Library

The Alternative Farming Systems Information Center (AFSIC), one of several centers at the National Agricultural Library (NAL) that provide in-depth coverage of specific agricultural subject areas, focuses on sustainable and alternative agricultural systems, crops and livestock.

AFSIC specializes in locating and accessing information related to many aspects of alternative agriculture: sustainable and organic crop and livestock farming systems; renewable farm energy options; alternative marketing practices; crop and livestock diversification including aquaculture, exotic and heritage farm animals, alternative and specialty crops, new uses for traditional crops, and crops grown for industrial production; and small farm issues.

INDEX

A

B

C

D

E

F

P

R

S

T

U

V

W

Y